HANDBOOK OF CITRUS BY-PRODUCTS AND PROCESSING TECHNOLOGY

HANDBOOK OF CITRUS BY-PRODUCTS AND PROCESSING TECHNOLOGY

ROBERT J. BRADDOCK
University of Florida
Institute of Food and Agricultural Sciences
Lake Alfred, Florida

A Wiley-Interscience Publication
JOHN WILEY & SONS, INC.
New York / Chichester / Weinheim / Brisbane / Singapore / Toronto

Florida Agricultural Experiment Station Journal Series No. R-067467.

For ordering and customer service, call 1-800-CALL-WILEY.

Library of Congress Cataloging-in-Publication Data

Braddock, Robert J. (Robert James), 1940–
    Handbook of citrus by-products and processing technology / Robert J. Braddock.
        p.   cm.
    Includes index.
    ISBN 0-471-19024-1 (cloth : alk. paper)
    1. Citrus fruit industry—By-products.   2. Citrus products.
  3. Citrus juices.   I. Title.
TP441.C5B73   1999
664'.804304—dc21   99-10821

Printed in the United States of America

10 9 8 7 6 5 4 3 2 1

CONTENTS

PREFACE

The importance of citrus in world commerce since the early 1900s is documented by hundreds of technical articles, texts, patents, and trade reports related to juice and product manufacture. Modern communications, consumer awareness of high-quality citrus, and product transportation methods have also resulted in significant production and processing growth emerging in countries with suitable climates for citrus culture. The large amount of fruit processed for juice recovery requires an understanding of modern technology and a large by-products industry to utilize the peel residue, essential oils, and other components that comprise almost half of the fruit mass. It is important for those planning to increase citrus production to deal with the issues of residue disposition because, inevitably, more fruit will be processed than sold fresh once regional production exceeds per capita consumption of whole fruit.

Fruit juice processing is generally not complicated but requires adherence to quality and sanitary principles through all phases of juice processing unit operations from fruit handling to product blending, packaging, and storage. For citrus, traditional methods of juice extraction and handling are still in vogue; however, the advent of newer separation technologies, automation, and aseptic bulk storage are fast being implemented in the industry. Some new technologies, which affect product chemistry, have resulted in product quality improvements. Others have been implemented for the purpose of improving process manufacturing efficiencies or decreasing waste. Knowledge of the basic composition of the fruit and its component parts and the relation to various processing techniques is essential to designing and applying new technology.

The traditional citrus by-products of cattle feed pellets and essential oils are still the most important utilization of juice processing residue, necessitated by the large mass available and the unique composition of soluble sugars, pectin, and volatile oils. As for many foods, world consumer interest in health benefits of naturally occurring chemical constituents has also stimulated investigations of certain compounds in citrus residues. Since the concentration of these compounds in the waste streams may be very low, it is imperative to not lose sight of the practical aspects of separation and recovery technologies. Also, economics of waste handling must be considered for any potential manufacture of these as new citrus by-products.

The approach of this text is to briefly recognize and describe the various traditional and modern citrus processing technologies, where they apply to juice and by-products. This book is intended as a resource for industrial and academic scientists, engineers, processors, and students searching for descriptions of current industry technology. There are many details and references included, without being tedious, to provide basic knowledge of the practical chemical aspects of juice products and their manufacture. Similarly, technological and chemical descriptions of the by-products are discussed. Finally, the author has learned much from his students, industrial colleagues, academic peers, and friends from many locations. It is hoped that this experience will be valuable and useful in the form of applications presented in this text.

ROBERT J. BRADDOCK

HANDBOOK OF CITRUS BY-PRODUCTS AND PROCESSING TECHNOLOGY

CHAPTER 1

THE CITRUS INDUSTRY

1.1 INTRODUCTION

The appeal of citrus fruits and products to consumers has resulted in expansion of production from origins in China to most of the world's subtropical and tropical areas. Strong demand has increased current world production by greater than 30% in the last 5 years, and this trend should continue into the near future (Florida Department of Agriculture, 1999). Production is principally oranges (43 million metric tons, mt), followed by mandarin hybrids and tangerines (>11 mt), grapefruit (4 mt), and lemons (>3 mt). Brazil and the United States are the largest orange producers, growing 60% of the world's oranges and processing 85% of their production. The United States, mostly Florida, is the largest grapefruit and lemon producer; China produces the most tangerines. Production growth in Florida will continue into the 2000s because newer plantings after the freezes of the 1980s are in areas south of severe freeze zones. Brazilian production is stabilizing, and there is some trend toward increasing utilization of fresh fruit, decreasing the processing percentage (Table 1.1). Significant production and processing growth are also emerging in the Mediterranean countries and China (Johnson, 1994). The large amounts of fruit being processed dictate evolution of a significant by-products industry to utilize the peel residue, essential oils, and other components as well as a search for newer products and uses.

1.2 HISTORICAL BASIS FOR PROCESSING

The traditional development of juice and products has followed the path of increasing production beyond the amount that can be sold successfully in the

TABLE 1.1 Brazilian Orange Production and Utilization

Season	Production (mt)	Fresh (mt)	Processed (mt)
1992–1993	12.8	1.6	11.2
1993–1994	12.5	2.3	10.2
1994–1995	12.7	2.6	10.1
1995–1996	14.1	4.1	10.0

Source: USDA/FAS, July 13, 1995, Agr. Attaché, São Paulo, Brazil.

fresh-fruit market. At first, only small canneries made juice from blemished fruit rejected as not meeting fresh-fruit quality standards. When the volume of fruit, juice, and residue is small, distribution of the juice is local and disposal of the residue is difficult. This disposal problem easily becomes apparent to even the smallest juice processor as piles of peel residue begin to sour and ferment. The magnitude of the problem of residue disposition must be considered by the planners of increasing citrus production because, inevitably, more fruit will be processed than sold fresh if production exceeds per capita consumption of whole fruit.

Industry factions in California and Florida early in the 1900s recognized the importance of processing technology and product development research to the growth of the citrus industry. During this time, annual Florida citrus production had again reached 5 million boxes (0.2 mt) (Florida Department of Agriculture, 1999) following the freezes of 1894–1895, and growers were concerned about the sale of future crops. Historical product research financed by the Florida Citrus Exchange in 1911 has been reviewed (Timmons, 1950). The relevance and insight of this early research to today's industry is apparent if one examines the notable findings as follows:

1. Orange and grapefruit juices could be preserved by pasteurization.
2. These juices were concentrated by vacuum distillation, after which the flavor of the concentrate was fortified by a citrus oil product.
3. Orange peel oil could be manufactured by vacuum steam distillation, although it was stated this would not be profitable for grapefruit because of the low quantity of oil in the peel.
4. The juice extraction wet residue was found to be suitable as a cattle feed.
5. Fermentation of waste juice, cull fruit, and other material to produce ethanol and citric acid was considered.

Timmons (1950) also stated several "firsts" for the Florida industry of those days. These included peeling with a 2–3% lye solution to produce grapefruit sections (1921), dehydration of several hundred pounds of grapefruit peel residue from a commercial cannery and experimental feeding to cattle at the University of Florida (1922), freeze concentration of orange juice (1920), hot-pack orange concentrate (1924), and use of a dairy evaporator to make commercial packs of concentrated orange juice (1925). The value of orange oil also was recognized very early, based on research to examine methods for commercial recovery and by comparing the quality of Florida oils with imported Sicilian oil (Hood, 1916).

There was also activity in the California processing industry during this same time period. A classic study documents the production in California of the following citrus by-products between 1899 and 1915: citric acid; cold-pressed, distilled, and terpeneless orange and lemon oils; pectin; dried peel; candied peel; fruit purees; preserved juices; syrup; alcohol; wine; and vinegar (Will, 1916). Will stated that this industry would become important because growers were not receiving adequate returns for their fruit and were beginning to realize the urgency for utilizing the waste fruit. For long-term success, he wrote that the industry must be based on cheap, accessible fruit, large-scale efficient processes and machinery, complete fruit utilization, good business management, ample capital, and a stable demand and value for the manufactured products. By the 1920s, products established in commercial markets included bottled juice, candied peel, vinegar, marmalades, jams, and jellies, and manufacturing methods were published (Chace, 1922). Certainly, one might learn from history, many of these early products and factors are still relevant to today's modern citrus processing and by-products industry.

As consumers demanded juices and concentrate, more fruit was processed and availability of by-products increased. The growth of the by-products industry was intimately bonded to the utilization of fruit for juice, and growth of the juice industry bonded to juice quality and availability. While canned juices and concentrates produced before 1940 had suitable commercial appeal, the availability of home refrigerators after 1945 allowed consumers to maintain products that were not commercially sterile. The advantage of manufacturing concentrate to reduce shipping expenses had long been known. However, early concentrated juices had poor flavor quality until the advent of technology to improve the flavor by adding some fresh juice back to the evaporator concentrate and maintaining the concentrate in the frozen state (MacDowell et al., 1948). This quality improvement rapidly increased consumer demand for juice and resulted in large-scale manufacture of juice, concentrate, and by-products.

The necessity and methods for dealing with waste from juice processing have been documented by more recent publications than those earlier mentioned. A U.S. Department of Agriculture (USDA) publication described the manufacture and utilization of most current citrus by-products during the first half-century of the industry in the United States (Chace et al., 1940). Manufacturing technology and properties of Florida citrus essential oils were first published after World War II (Kesterson and McDuff, 1948) and later (Kesterson et al., 1971). Peel residue and juice plant waste material was established as a suitable cattle feed supplement in the 1920s (Mead and Guilbert, 1926, 1927) and as a major industry by-product by the 1950s (von Loesecke, 1950). A number of other publications have documented the continuing development of the citrus processing and by-products industry (Braverman, 1949; Kesterson and Hendrickson, 1958; Sinclair, 1961; USDA, 1962; Kesterson and Braddock, 1976; Braddock and Cadwallader, 1992; Braddock, 1995). General citrus processing technology also has been the subject of several excellent books published in the last 25 years (Sinclair, 1972, 1984; Nagy et al., 1977). More recent general fruit juice processing texts with citrus chapters have been published (Nagy et al., 1993; Somogyi et al., 1996). The practical, applied aspects of citrus processing in a manufacturing environment, including product quality control, processing technology, and by-products information, have also been the subject of a three-volume set of texts (Redd et al., 1986, 1992, 1996). A unique book that includes descriptions of processing quality control, personnel management methods, and a number of basic computer programs for blending and mixing juices and concentrates is available (Kimball, 1991). Kimball has also written about the manufacturing and technological aspects of handling and processing oranges and tangerines (orange-colored fruit) and grapefruit, lemons, and limes (yellow fruit) into juices and by-products (Kimball, 1996). A text useful for those interested in following increases or decreases in fruit production due to climatic freezes over the last century has also been published (Attaway, 1997).

REFERENCES

Attaway, J. A. (1997). *A History of Florida Citrus Freezes*, Florida Science Source, Inc. Lake Alfred, FL.

Braddock R. J. (1995). By-products of citrus fruit, *Food Technol.* **49**(9):74, 76–77.

Braddock, R. J. and Cadwallader, K. R. (1992). Citrus by-products manufacture for food use, *Food Technol.* **46**(2):105–110.

Braverman, J. B. S. (1949). *Citrus Products*; *Chemical Composition and Chemical Technology*, Interscience, New York.

Chace, E. M. (1922). By-products from citrus fruits, USDA Cir. No. 232, Washington, D.C.

Chace, E. M., von Loesecke, H. W., and Heid, J. L. (1940). Citrus fruit products, USDA Cir. No. 577, Washington, D.C.

Florida Department of Agriculture (1999). Citrus summary. 1997–98, Florida Agr. Statistics Service, Orlando, FL.

Hood, S. C. (1916). The production of sweet-orange oil and a new machine for peeling citrus fruits. Possibility of the commercial production of sweet-orange oil from waste oranges, USDA Bull. No. 399, Washington, D.C.

Johnson, T. M. (1994). Trends in citrus juice processing, *Citrus Ind. Mag.* **75**(6):42, 45, 48, 50.

Kesterson, J. W. and Braddock, R. J. (1976). By-products and specialty products of Florida citrus, Fla. Agr. Exp. Sta. Tech. Bull. No. 784, University of Florida. Gainesville, FL.

Kesterson, J. W. and Hendrickson, R. (1958). Utilization of citrus by-products, *Econ. Bot.* **12**(2):164–185.

Kesterson, J. W. and McDuff, O. R. (1948). Florida citrus oils: Commercial production and properties (1947–48 season), Fla. Agr. Exp. Sta. Tech. Bull. No. 452, University of Florida, Gainesville, FL.

Kesterson, J. W., Hendrickson, R., and Braddock, R. J. (1971). Florida citrus oils, Fla. Agr. Exp. Sta. Tech. Bull. No. 749, University of Florida, Gainesville, FL.

Kimball, D. A. (1991). *Citrus Processing Quality Control and Technology*, Van Nostrand Reinhold, New York.

Kimball, D. A. (1996). "Oranges and tangerines" (Chap. 9); "Grapefruits, lemons and limes" (Chap. 10). In *Processing Fruits: Science and Technology*, Vol. 2, *Major Processed Products*, L. L. Somogyi, D. M. Barrett, and Y. H. Hui, Eds., Technomic Publishing, Lancaster, PA, pp. 265–304, 305–336.

MacDowell, L. G., Moore, E. L., and Atkins, C. D. (1948). Method of preparing full-flavored fruit juice concentrates, U.S. Pat. 2,453,109.

Mead, S. W. and Guilbert, H. R. (1926). The digestibility of certain fruit by-products as determined for ruminants. Part I. Dried orange pulp and raisin pulp, Agr. Exp. Sta. Bull. No. 409, University of California, Berkeley, CA.

Mead, S. W. and Guilbert, H. R. (1927). The digestibility of certain fruit by-products as determined for ruminants. Part II. Dried pineapple pulp, dried lemon pulp, and dried olive pulp, Agr. Exp. Sta. Bull. No. 439, University of California, Berkeley, CA.

Nagy, S., Chen, C. S., and Shaw, P. E. (1993). *Fruit Juice Processing Technology*, AgScience, Auburndale, FL.

Nagy, S., Shaw, P. E., and Veldhuis, M. K. (1977). *Citrus Science and Technology*, Vols. 1 and 2, AVI, Westport, CT.

Redd, J. B., Hendrix, D. L., and Hendrix, C. M., Jr. (1986). *Quality Control Manual for Citrus Processing Plants*, Vol I, Intercit, Safety Harbor, FL.

Redd, J. B., Hendrix, D. L., and Hendrix, C. M., Jr. (1992), *Quality Control Manual for Citrus Processing Plants*, Vol. II, AgScience, Auburndale, FL.

Redd, J. B., Shaw, P. E., Hendrix, C. M., Jr., and Hendrix, D. L. (1996). *Quality Control Manual for Citrus Processing Plants*, Vol. III, AgScience, Auburndale, FL.

Sinclair, W. B. (1961). *The Orange: Its Biochemistry and Physiology*, University of California Press, Berkeley, CA.

Sinclair, W. B. (1972). "Grapefruit by-products." In *The Grapefruit*, University of California Press, Riverside, CA, Chap. 6, pp. 502–551.

Sinclair, W. B. (1984). *The Biochemistry and Physiology of the Lemon and Other Citrus Fruits*, University of California Press, Oakland, CA.

Somogyi, L. P., Barrett, D. M., and Hui, Y. H. (1996). *Processing Fruits: Science and Technology*, Vol. 2, *Major Processed Products*, Technomic, Lancaster, PA.

Timmons, D. E. (1950). Citrus canning in Florida: Early history and current statistics, AE Series 50-4, January, Fla. Agr. Exp. Sta., University of Florida, Gainesville, FL.

USDA. (1962). *Chemistry and Technology of Citrus, Citrus Products, and By-products*, Agriculture Handbook No. 98, USDA, Washington, D.C.

von Loesecke, H. W. (1950). Citrus cannery waste, its use and disposition, U.S. Bur. Agr. & Indus. Chem. AIC #290, November, USDA, Washington, D.C.

Will, R. T. (1916). Some phases of the citrus by-product industry in California, *J. Ind. Eng. Chem.* **8**(1):78–86.

CHAPTER 2

FRUIT GROWING, HARVESTING, HANDLING, AND GRADING FOR PROCESSING

Citrus fruit production requires climatic conditions found in warmer climates from approximately 40°N to 40°S of the world citrus belt. The main planting areas within this belt are 20°–40°N and 20°–40°S latitudes. Characteristics of these main planting areas, which are most suitable for producing good-quality fruit, are definite seasonal changes and cool nights, possibility of periodic frosts and freezes, rainfall/drought, and good natural fruit color, sweetness, and tartness. Longer maturation times are necessary in the main planting latitudes. For many varieties of citrus, the fruit will mature from 7 (lemons, limes) to 10 months (oranges, grapefruit, tangerines) after tree flowering. Valencia oranges are unusual in that maturation can be greater than 12 months after flowering, resulting in slightly different cultural and harvesting practices. Characteristics of the growing areas near the equator are constant warm day/night and seasonal temperatures, rainfall without drought, rapid fruit maturation and short harvesting periods, poor fruit and juice color development, lower soluble solids, and much lower acid resulting in high °Brix/acid ratios (see Chap. 3). These latter factors cause the juice to have an insipid, watery flavor.

2.1 CULTURAL PRACTICES AND COSTS

Citrus trees are generally propagated by grafting (budding) a twig or bud from the target tree to rootstocks grown from seeds, conferring desirable traits (drought resistance, cold hardiness, fruit quality, etc.) of the rootstock to the tree. The seedlings are grown in a nursery and planted in a commercial grove at a minimum age of 12–15 months to a maximum of about 2 years of age.

7

Planting density is currently about 150 trees/acre (370 trees/ha), although trends are to higher planting densities. Tree costs at this density in Florida may amount to $2.50–$3.50/tree, planting labor is $1.50/tree, tree wrap freeze protection is $2.00/tree; and labor and materials for the first planting year are $3.00/tree. The sum of these costs may amount to $1013/acre for the first year (labor portion = $633/acre). A grower of fruit for processing ($1.20/lb juice soluble solids content) may expect the first profit 6–7 years after planting (10–12 years if land must be purchased). The greatest costs for fruit production are for pest control sprays, herbicides, irrigation, and fertilizer. Thus, some savings may be possible growing fruit that is not required to be blemish-free for processing. A typical example (Table 2.1) of the costs for Florida Valencia oranges for processing includes both cultural and harvesting costs. As one can determine, the yields of fruit/acre and juice soluble solids yield can affect the total fruit cost (Muraro, 1998).

Future world industry cultural practices are directed toward increasing growing efficiency and fruit cost reduction. Low-cost fruit is especially necessary for processing fruit because processing costs added to production and harvesting costs determine profitability of the finished juice and by-products. The cost relationships between processors' needs for quality fruit and growers' desire for maximum returns have to be balanced for both long- and short-term goals. For example, tangerines, Temples, and murcotts are valuable to the grower as fresh fruit, but not for processing. The desirable high color of the juice does not offset the distinct off-flavors and bitterness, decreasing value to the processor (Williard and Hendrix, 1996).

Higher fruit and juice yields increase the return on investment costs for both the processor and grower. Also, juice with higher sugar concentration improves grower profit and reduces the processors' costs. This is especially true when water has to be evaporated to make concentrated juice. In Florida, growers of

TABLE 2.1 Delivered-in Costs for Central Florida Valencia Oranges for Low-Cost Processed Cultural Program

	$/Acre[a]	$/Box[a]	$/lb Solids[a]
Cultural costs	766	1.77	0.27
Interest on cultural costs	38	0.09	0.01
Management	48	0.11	0.02
Taxes/regulatory	59	0.14	0.02
Total direct grower costs	911	2.10	0.32
Interest on capital investment	376	0.87	0.13
Total grower costs	1287	2.97	0.46
Harvesting/assessment	877	2.03	0.31
Total delivered-in costs	2164	5.00	0.77

Source: Muraro (1998).

[a]433 boxes/acre, 6.5 lb solids/box, 102 trees/acre.

TABLE 2.2   Fruit, Juice, and Soluble Solids Yield of Florida Citrus Fruit for Processing

Fruit	Box/Acre	ps[a]/Box	ps/Acre
Early-mid[b]	460	6.0	2760
Valencia[c]	405	6.5	2633
Temple[d]	330	6.0	1980
White grapefruit[e]	455	4.5	2043
Red grapefruit[b]	500	4.5	2261

[a]ps = pounds of soluble solids based on sucrose.
[b]Southwest Florida average.
[c]Central Florida average.
[d]Statewide average.
[e]Indian River average.

fruit for processing are paid on the basis of juice yield and sugar content (pounds soluble solids, ps). Some of the common fruits grown for processing are compared on a ps/acre basis in Table 2.2. Higher fruit yields of modern cultural practices and younger plantings are reflected in the values of Table 2.2, with the exception of Temple oranges. Data for Temples reflect state averages of young and old plantings and a wide variety of cultural practices. For the future, modern efficient practices will be required for maximum profitability.

Future new citrus cultivars from genetic breeding selections must consider such factors as juice quality, yield, and season of maturity to meet processors' as well as growers' requirements (Gmitter, 1995). Cultural trends include higher density plantings, which may increase tree count to over 300 trees/acre. Greater tree density has the advantage of more fruit/acre in a shorter time. Disadvantages are that initial planting and cultural costs are higher and different grove management practices are necessary. Tree size control by either hedging or genetics is a trend that affects ease of grove maintenance, harvesting, and fruit yield. Integrated pest control, reduced fertilizer application rates, and low-rate chemical control of weeds are trends that can reduce costs. Water management, conservation practices, and general environmental awareness are also important to the future of the industry. On a world scale, larger unit size for management is also a trend, which reduces costs. As these trends become practice and citrus production increases, new efficient processing plants must be built to utilize the fruit.

2.2   HARVESTING

Operation of citrus processing plants is dependent on fruit availability throughout the growing season. It is beyond the scope of this book to describe the seasonal maturation of all the world's citrus varieties and locations; however, examples of the two primary processing locations (Florida, Northern Hemisphere; Brazil, Southern Hemisphere) are presented in Table 2.3. The seasons for the different varieties overlap somewhat for each location, allowing citrus processing plants to operate for 8–9 months each year. This length of time is one thing that makes citrus processing unique among all the world's fruit because most other fruit have very short harvest seasons. Also, it should be noted that the most intense processing occurs for Hamlin through Valencia seasons, with the Northern and Southern hemispheres approximately 6 months out of phase with each other.

Citrus fruits have traditionally been harvested by hand-picking, and this is likely to continue into the future in parts of the world where labor is cheap and manpower is adequate. Fruit for the fresh market is picked and put in a bag carried by the picker, who empties the bag into a larger machine-transportable container. After 8–10 years of age, most trees are large enough to require use of a ladder to reach the fruit, stimulating interest in tree size research mentioned above. Processing fruit may be harvested and handled entirely differently. In many cases, pickers climb trees and shake limbs until fruit falls on the ground, then it is picked up and put into large tubs or containers,

TABLE 2.3   Fruit Variety and Months of Maturity for Processing[a]

Florida[b]	Brazil[c]
J F M A M J J A S O N D	J F M A M J J A S O N D

Florida[b]

```
                HAML
                ← →
  ← →           AMBS
                ← →
                TANG
                ← →
  PINE
  ← →
  TANL
  ← →
  TEMP
  ← →
  GPFT          ← →
  ← →
       VALN
       ← →
          LEMN
          ← →
          LIME
          ← →
```

Brazil[c]

```
                HAML
                ← →
                PERA
                ← →
                     VALN
                     ← →
                     MURC
                     LEMN
                     ← →
  LIME
  ← →
                     GPFT
                     ← →
```

[a]HAML, Hamlin; AMBS, Ambersweet; TANG, tangerine; PINE, pineapple orange; TANL, tangelo; TEMP, Temple; GPFT, grapefruit; VALN, Valencia; LEMN, lemon; LIME, Persian lime; PERA, Pera orange; MURC, Murcott.
[b]Florida Department Agriculture (1996).
[c]Figueiredo (1991).

which may hold more than 1000 lb of fruit. These tubs are transported to a location near the grove where the fruit is dumped into large truck-trailers and hauled to the processing plant. These trailers hold from 45,000 to 50,000 lb of fruit. In some citrus areas, both larger, tandem-trailer and/or much smaller trucks may be used.

Since most pickers are paid on the basis of volume of fruit picked, whatever is on the ground under the trees, including drops and spoiled fruit, ends up in the tubs destined for the processing plant. Amounts of trash, sand, and other debris are recognized sanitation and cleaning problems at the plant, and harvesting quality improvement programs are in place in the industry. The United States and other areas where hand harvesting labor can be unreliable or expensive have current interest in mechanical fruit harvesting. In Florida, peak interest in developing machine harvesters was during the mid-1970s, during a period of hand labor shortage. It is felt that mechanical harvesting costs will eventually approach economic competitiveness. Thus, there is a continuing harvesting program in place in Florida with the goal of ensuring harvest of future crops at competitive costs. Florida industry predictions for the next decade include increasingly larger crops, low fruit prices, increasing harvest costs, and more government labor regulations. Planners are looking to improve machine design, examine tree size and configuration, reexamine use of abscission chemicals, and coordinate academic and industrial research. For the interested reader, a description of harvester types and important factors related to mechanical harvesting of citrus has been presented in detail (Whitney, 1995).

2.3 HANDLING AND GRADING

2.3.1 Fresh Market Fruit

Harvesting and handling fruit for the fresh market has some impact on processing because fruit not meeting grades and quality standards may be destined for a processing plant. Fresh-fruit practices that affect processing will be described briefly in the following statements. The additional handling and storage required in a packinghouse may reduce juice and peel oil yields because bruising and moisture loss causes physical softening of peel and internal tissues, reducing turgidity of oil glands and juice vesicles. The time frame between harvest and elimination by grading at a packinghouse may be from one day to several weeks before arrival at a processing plant. For lemons, which are degreened (cured) under controlled temperature conditions, several gradings over many weeks may take place. Since many packinghouses are much smaller than processing plants, a trailer for eliminations may take a day or so to fill, allowing the fruit to sit in the sun. Waxing and anaerobic storage conditions also affect fruit metabolism, increasing concentrations of ethanol, acetaldehyde, and other volatiles (Baldwin et al., 1995). Other studies also have discussed the decreased sensory and chemical attributes of juice from fruit during storage and handling

(Baldwin, 1993; Ke and Kader, 1990; Cohen et al., 1990). These factors indicate that fruit from packinghouse eliminations may have poor juice quality.

There has been some recent resistance to processing eliminations because certain postharvest fungicides, such as thiabendazole, have been detected in packs of commercial citrus juices and concentrates (Bushway et al., 1995). It is a matter of conjecture whether or not there is a health concern from consumption of such products, since parts per billion (ppb) and smaller concentrations of chemical residues may now be detected in many foods. However, there is concern in some European countries and Japan about the presence of these chemicals in juice. From the perspective of the citrus processor, only a very small proportion of the total fruit processed would come from packinghouse eliminations. In the large processing areas of Brazil and Florida, most oranges are grown and harvested specifically for processing and never come in contact with postharvest fungicides. The rational mind might reason as follows: For oranges in Florida, 94% are utilized for processing, leaving 6% handled for the fresh-fruit market. Certainly the percentage treated with fungicide that is graded out to processing is very small, causing extreme dilution in the total juice stream. The situation is similar with grapefruit, except that the processing/fresh ratio is lower, 60/40% (Florida Department of Agriculture, 1996). Also, many processing plants in Florida and Brazil do not process eliminations because they are unaffiliated with any fresh-fruit packinghouses.

There are two technological developments in handling fresh fruit that could have an impact on processing. The first of these, high-pressure washing, may have some advantages for better fruit surface cleaning and sanitizing as well as reduction of water usage, resulting in increased packinghouse production efficiency (Katz, 1995). This technique is receiving some interest from unpasteurized, chilled-juice processors to reduce microbial loads on fruit and equipment. The other technology—automatic grading for color and blemishes, sizing, and some internal disorders—is in commercial application in fresh-fruit houses (CVS, 1998). Current status makes use of video cameras, ultrasonic sizing, X-ray, or load cell units for lemons, oranges, and grapefruit (Miller, 1987). Beyond fruit density determinations, internal quality measurements are not yet reliable. Disadvantages of this technology for processing are that systems are too slow at present, <500 boxes/hr, compared to 5000 boxes/hr total rates in large processing plants. Advantages for the fresh-fruit industry are attained through more grading consistency, labor reduction, and identification with a premium product (ASAE, 1994).

2.3.2 Fruit Receiving and Grading for Processing

Fruit for processing is generally grown, harvested, and handled with intentions of juice manufacturing. Before harvest, fruit must reach some level of maturity to assure quality products after processing. A grower determines maturity, color break, and juice content before deciding to harvest. In Florida, there are state governmental regulations and maturity tests for processed fruit that must be met, partly to assure the grower the proper value for the fruit, but also to assure that the consumer receives a quality product. The fruit for these tests comes randomly from each trailer load after weigh-in during unloading and during the initial grading at the processing plants. These tests have been discussed elsewhere (Miller and Hendrix, 1996; Kimball, 1991) and the regulations are available (Florida Department of Citrus, 1996). The Florida state test procedures are required by law and a fee is charged the processors. There is a marketing agreement between the U.S. Department of Agriculture, the Florida Department of Agriculture and Consumer Services (FDOACS), growers, processors, and shippers to perform these tests and provide inspectors at each plant. This is a formal procedure with the objective of providing an unbiased third party to assure quality of fruit and products. An upgrade of the FDOACS system for new electronics hardware, communications, and software is currently being implemented in the Florida industry (Ratajeski, 1994). The analytical procedures are automated and determine juice yield, °Brix, acid, and °Brix/acid ratio. A computer calculates the results of the tests, prints useful information (e.g., ps/box), and both the processor and grower receive a copy. An automated test/extraction procedure eliminates human error and expedites payment for the fruit.

For protection of the processor and fruit dealer, some accounting procedure should be in place. In Florida, each trailer load is accompanied by an FDOACS trip ticket, which is presented during fruit receiving. The fruit and trailer are weighed, unloaded, and the empty trailer weighed again. An average load weight is about 45,000 lb, or 500 boxes. Unloading is done by dumping fruit into a receiving conveyor from inclined ramps or hydraulic lifts. Unloading is very fast at large processing plants and can be completed in less than 10–15 min. The regulations govern the amount of culls and unwholesome fruit allowed in a load, and this amount is subtracted from the weight for payment.

2.3.3 Bin Handling and Storage

Most processing plants around the world unload and receive the fruit dry. There have been some commercial attempts to do wet receiving, where advantages were stated to be low maintenance, safety, reliability, long life, and cleaner fruit, making grading easier (Curls, 1970). Disadvantages found from commercial practice are that water costs are high, the effluent gets very dirty and must be monitored for chemical contaminants before discharge, and fruit pump maintenance may be more than dry line conveyors. In dry unloading and conveying systems, a grading step for elimination of trash, culls, drops, splits, sand, and leaves is necessary (Fig. 2.1). Trash elimination roller conveyors and mechanical destemmers are required to remove sand, leaves, twigs, and small branches (Kimball, 1996). Fruit from mechanical harvesting operations is much more difficult to clean, and there are efforts to increase research in this field (Whitney, 1995). It has been estimated that up to 200 lb of trash per 500-box load are received (Miller and Hendrix, 1996). In most of the world, fruit bin

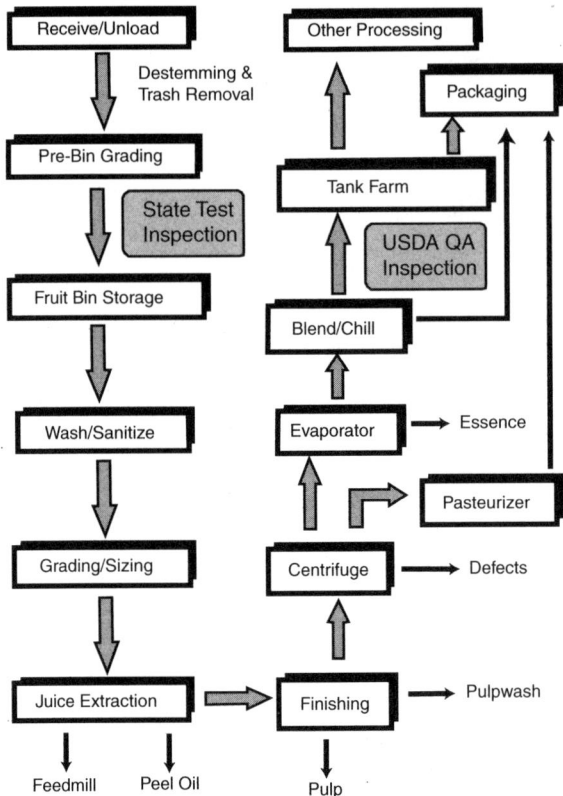

Figure 2.1   Citrus plant flow diagram for fruit handling and processing operations. Peel oil is recovered between the washing and grading operations in the Brown extractor process (see Chap. 11).

storage is common after unloading and trash elimination. It has also been common to wash fruit before grading and bin storage. In recent years, many processors have discontinued washing or rinsing fruit before bin storage because sending wet fruit to the bins can increase microbial loads downstream, making sanitation more difficult.

Prebin grading for fruit maturity and unwholesomeness is good manufacturing practice and should be performed. As stated, grading is required by law in Florida and assures better quality fruit in the bins. A load of fruit will be rejected if the maximum unwholesome fruit count is >10%, including 2% decay (Florida Department of Citrus, 1996). Some processors use mechanical grading methods such as inclined ramps and belts where, in principle, a sound fruit will roll faster and drop farther from the end of the ramp than an unsound fruit (Bryan et al., 1978; Miller and Hendrix, 1996). As each trailer is unloaded, a statistical sample representative of the load is taken automatically by a computerized singulator/mechanical arm device that can determine how much fruit to take and the destination of the samples. This sample represents about 1 lb of fruit per 10 boxes in the load, or approximately 140–150 oranges. These samples are used by the processor to measure the juice yield and quality parameters mentioned above. Where this process is automated and computer controlled, data maintenance and record keeping is simplified and of tremendous advantage for the processor for comparison purposes in the operations of the plant.

2.3.4 Direct Unloading

Recent modernization of the fruit unloading process has led to the concept in Florida and Brazil referred to as direct unloading. Simply stated, this means that fruit goes directly from unloading and grading to the juice extraction operation, bypassing or eliminating the storage bins. One variation of the technique in Brazil utilizes a metal box used as a grove container for the fruit, which is lifted onto a flat-bed truck and hauled to the processing plant, where it serves as a mobile bin (Carvalho, 1993). In Florida, as in Brazil, using direct unloading reduces the number of times fruit is dropped or bruised and has advantages of increased juice yield and improved juice quality (Mitenius, 1996). As shown in Table 2.4, there are also some economic advantages in capital and operating costs compared with conventional bin storage systems. Such systems require careful attention to coordinate fruit procurement and harvesting schedules with plant operation and blending for quality. There is an additional investment requirement for more fruit trailers and increased plant yard space for trailers waiting to unload.

2.3.5 Surge Bin

Whether or not a plant stores fruit in bins or uses direct unloading, there have to be operator stations for sequential start and stop of belts, conveyors, and

TABLE 2.4 Direct Unloading vs. Bin Costs for Processing 4 Million Boxes/Year

	Bin Storage	Direct Unloading
Capacity (boxes)	10,000	10,000
Bin units	24	75
Trucks	50[a]	12[b]
Bin cost/unit ($)	18000	6150
Total investment ($)	2,057,000	1,421,000
Operation cost ($/yr)	1,100,000	640,000

Source: Carvalho (1993).

[a]50% rental trucks.
[b]Three trucks for plant unloading.

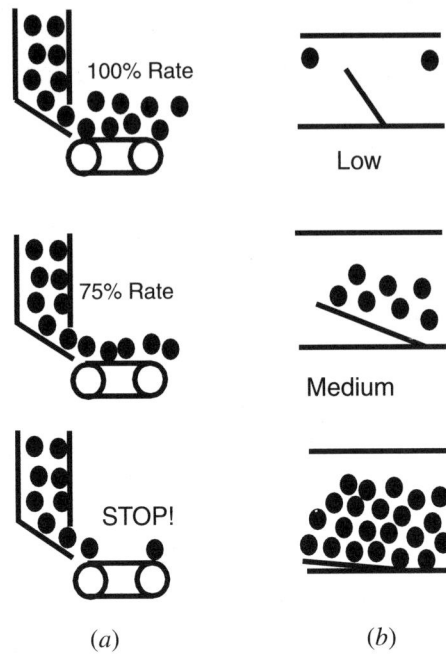

(a)   (b)

Figure 2.2   Fruit flow rates for operation of the (a) surge bin and (b) return belts.

elevators needed to move fruit. A fruit surge bin is the critical control point required to maintain a steady fruit supply for the juice extraction operation. The surge bin affects grading, juice yield, quality, finisher operation, and other downstream plant operations. This system is automated in many cases and controls the operations from trailer unloading to bin filling and from fruit release to filling the surge bin. The surge bin fruit level is kept above the exit gate with an infrared detector, strain gage, or other control method. If fruit is below the set-point level, fruit supply to the surge bin starts automatically. Fruit rate to the juice extractors is controlled by speed of the pull-out belt. Initially, the rate is steady, depending on the juice flow. Extractor return fruit is monitored, and, if excessive, the surge bin pull-out belt is slowed, which prevents graders and mechanical sizers before the extractors from being overloaded. Figure 2.2 illustrates how fruit flow to the juice extractors is controlled by the surge bin and return fruit belts. The object is to maintain a steady fruit flow at 100% of the optimum rate, with few fruit passing the extractor line to the return belt.

2.3.6   Postbin Handling

After bin storage and before juice extraction, fruit washing, sanitizing, and final grading are the most important operations. Proper washing ensures the fruit is clean, improves grading efficiency, and is very necessary for juice quality. It is the graders' responsibility to decrease the amount of culls and unwholesome fruit (Kimball, 1996; Miller and Hendrix, 1996). Sanitizing is essential for control of spoilage microbes, which may contaminate conveying equipment and juice extractors and affect juice quality. Applied citrus plant sanitation requirements have been published (Redd et al., 1992; Winniczuk, 1994). Citrus microbiology and processing plant fruit sanitation is important to maintain juice quality and prevent product losses from spoilage (Kimball, 1991). Some smaller processors are marketing nonpasteurized, chilled, and/or frozen single-strength fresh juice, which has limited shelf life and distribution. There has been at least one documented outbreak of human illness caused by persons consuming contaminated commercially packed fresh nonpasteurized citrus juice (Parish, 1998). This possibility requires careful attention to fruit and equipment sanitation in the case of nonpasteurized juice and also has serious implications for large processors (Schrader and Kane, 1996).

Careful cleaning and rinsing of equipment is a sanitary requirement; however, the practice of rinsing fruit with evaporator condensate water recycled through a cooling tower may actually contaminate fruit. Heat- and acid-tolerant bacterial spores from common soil genera, *Alicyclobacillus*, are present in condensate water and may contaminate fruit during rinsing. The spores and organisms have been isolated from concentrated juices and can cause product losses and flavor problems during storage and distribution. The spores are very heat resistant, with D values between 15 and 30 min at 90°C, and can attach to equipment and tank surfaces, resisting washing and rinsing (Wisse and Parish, 1998). (For a discussion of D values see Chapter 5.) It is suggested that careful attention to cooling towers and condensate reuse systems can minimize this microbial problem.

2.3.7 Sizing

Fruit sizing for juice extraction is necessary for proper operation of the extractors. Processing plant mechanical sizers allow conveyed fruit to pass through rollers or openings dependent on the fruit diameter and the sizer clearance. Sizer adjustment is possible during operation to allow for variations in the incoming fruit streams. Operation and organization of sizers and extractor lines is such that small fruits fall out first. Since variation in fruit diameter and peel thickness is great from small to large fruits, one extractor setup cannot handle all variations. Most juice extractors will be set up with cups near 3 inches because 80% of all oranges are within 0.5 inch of 2.875 inches diameter. Smaller fruits have more juice/fruit mass, and the juice has a higher Brix than large fruits. This allows some juice blending opportunities, provided that proper sizing is performed. It also should be mentioned that growing conditions (temperature, rainfall, etc.) and placement on a tree can affect the average seasonal fruit size and quality (Sites and Reitz, 1950). A prolonged drought might result in very small fruit for a season, for example. Obviously, a close match between extractor cup diameter and fruit size is advantageous and will improve juice yield and quality.

REFERENCES

ASAE (1994). Nondestructive technologies for quality evaluation of fruits and vegetables, Publ 05-94, Am. Soc. Agric. Engrs., St. Joseph, MI 49085.

Baldwin, E. A. (1993). "Citrus fruit." *Biochemistry of Fruit Ripening*. G. B. Seymour, J. E. Taylor, and G. A. Tucker, Eds., Chapman & Hall, Chap. 4, pp. 107–149.

Baldwin, E. A., Nisperos-Carriedo, M., Shaw, P. E., and Burns, J. K. (1995). Effect of coatings and prolonged storage conditions on fresh orange flavor volatiles, degrees brix, and ascorbic acid levels, *J. Agric. Food Chem.* **43**:1321–1331.

Bryan, W. L., Anderson, B. J., and Miller, J. M. (1978). Mechanically assisted grading of oranges for processing, *Trans. ASAE* **21**(6):1226–1231.

Bushway, R. J., Brandon, D. L., Bates, A. H., Li, L., Larkin, K. A., and Young, B. S. (1995). Quantitative determination of thiabendazole in fruit juices and bulk juice concentrates using a thiabendazole monoclonal antibody. *J. Agric. Food Chem.* **43**: 1407–1412.

Carvalho, W. M. de. (1993). *New Developments in Citrus Processing in Brazil. Citrus Processing: New Developments*, Proc. Food Industry Short Course, R. F. Matthews, Ed., University of Florida, IFAS, FSHN Dept., pp. 253–261.

Cohen, E., Shalom, Y., and Rosenberger, I. (1990). Postharvest ethanol buildup and off-flavor in "Murcott" tangerine fruits, *J. Am. Soc. Hort. Sci.* **115**(5):775–778.

Curls, A. (1970). Hydraulic conveyance of citrus fruit, *Trans. Citrus Eng. Conf. ASME* **16**:14–24.

CVS (1998). Automatic Blemish Sorter, www.cvsusa.com.

Figueiredo, J. O. de. (1991). "Variedades-copa de valor comercial." In *Citricultura Brazileira*, 2nd ed., Vol. 1, O. Rodriguez, F. Viejas, J. Pooldeu, and F. Amaro, Eds., Fundução Cargill, Campinas, São Paulo, Brazil, Chap. 10, pp. 241–278.

Florida Department of Agriculture (1996). Citrus summary, 1995–96, Florida Agr. Statistics Serv, Orlando, FL.

Florida Department of Citrus (1996). Official rules affecting the Florida citrus industry, Ch. 601, Florida Statutes, Part 3, Rules applying to processed products.

Gmitter, F. G., Jr. (1995). *Characteristics and Potential of New Citrus Cultivars*, Proc. Citrus Processing Short Course, C. A. Sims, Ed., University of Florida, IFAS, FSHN Dept., pp. 165–172.

Katz, M. (1995). High pressure washing boosts packout, *Citrograph*. **80**(8):3, 6–7.

Ke, D. and Kader, A. A. (1990). Tolerance of Valencia oranges to controlled atmospheres as determined by physiological responses and quality attributes, *J. Am. Soc. Hort. Sci.* **115**(5):779–783.

Kimball, D. A. (1991). *Citrus Processing Quality Control and Technology*, Van Nostrand Reinhold, New York.

Kimball, D. A. (1996). "Oranges and tangerines." In *Processing Fruits: Science and Technology*, Vol. 2, *Major Processed Products*. L. L. Somogyi, D. M. Barrett, and Y. H. Hui, Eds., Technomic, Lancaster, PA, Chap. 9, pp. 265–304.

Miller, W. M. (1987). Automated inspection/classification of fruits and vegetables, *Trans. Citrus Eng. Conf. ASME* **33**:42–52.

Miller, W. M. and Hendrix, C. M., Jr. (1996). "Fruit quality, inspection, handling, sampling and evaluation. In *Quality Control Manual for Citrus Processing Plants*, Vol. 3, J. B. Redd, P. E. Shaw, C. M. Hendrix, Jr., and D. L. Hendrix, Eds., Ag-Science, Auburndale, FL, Chap. 7, pp. 233–251.

Mitenius, N. (1996). *Direct Unloading Fruit Receiving*, Proc. Citrus Processing Short Course, C. A. Sims, Ed., University of Florida, IFAS, FSHN Dept., pp. 101–110.

Muraro, R. P. (1998). 1997–98. Comparative citrus budgets, *Citrus Ind. Mag.* **79**(July): 21–32.

Parish, M. E. (1998). Coliforms, *Escherichia coli* and *Salmonella* serovars associated with a citrus processing facility in a Salmonellosis outbreak, *J. Food Protect.* **61**(3): 280–284.

Ratajeski, R. (1994). New generation of state test house equipment, *Trans. Citrus Eng. Conf. ASME* **40**:1–12.

Redd, J. B., Hendrix, D. L., and Hendrix, C. M., Jr. (1992). "Plant operations: Quality assurance and sanitation." In *Quality Control Manual for Citrus Processing Plants*, Vol. 2, AgScience, Auburndale, FL, pp. 3–70.

Schrader, G. W. and Kane, L. (1996). Best practices for fresh juice production, *Trans. Citrus Eng. Conf. ASME* **42**:1–15.

Sites, J. W. and Reitz, H. J. (1950). The variation in individual Valencia oranges from different locations on the tree as a guide to sampling methods and spot-picking for quality. Part II. Titratable acid and the souble solids/titratable acid ratio of the juice, *Proc. Am. Soc. Hort. Soc.* **55**:73–80.

Whitney, J. D. (1995). A review of citrus harvesting in Florida, *Trans. Citrus Eng. Conf. ASME* **41**:33–59.

Williard, R. P. and Hendrix, C. M., Jr. (1996). "Factors that influence growers' returns at processing plants." In *Quality Control Manual for Citrus Processing Plants*, Vol. 3, J. B. Redd, P. E. Shaw, C. M. Hendrix, Jr., and D. L. Hendrix, Eds., AgScience, Auburndale, FL, Chap. 8, pp. 252–256.

Winniczuk, P. P. (1994). Effects of sanitizing compounds on the microflora of orange fruit surfaces and orange juice, Thesis, Citrus Research & Education Center, University of Florida, Lake Alfred, FL.

Wisse, C. A. and Parish, M. E. (1998). Isolation and enumeration of sporeforming, thermoacidophilic, rod-shaped bacteria from citrus processing environments, *Dairy, Food Env. Sanitation* **18**(8):504–509.

CHAPTER 3

COMPOSITION, PROPERTIES, AND EVALUATION OF FRUIT COMPONENTS

Chemical composition of citrus fruit components has been described extensively in the literature. For detail, the interested reader should consult one of the published texts (Kefford and Chandler, 1970; Nagy et al., 1977; Sinclair, 1972, 1984) or reviews of the subject (Ranganna et al., 1983). Rather than discuss the analytical chemistry and composition of citrus fruits, this chapter will consider the technology of manufacturing products from various citrus fruit component raw materials. Certain properties and compositional information, as well as mass balances of components, of common fruit varieties will be described.

3.1 PROPERTIES

3.1.1 Fruit Shape

The shape of citrus fruit is important for designing harvesting, handling, conveying, automatic grading, and other processing systems. Shapes and formulas for calculation of the volumes and surface areas of the various citrus fruits are illustrated in Figure 3.1. Oranges generally are considered spherical (equatorial diameter, $2b = 5.7 - 9.5$ cm), lemons ($2b = 4.4 - 6.4$ cm) and limes ($2b = 3.8 - 5.0$ cm) prolate, and grapefruit ($2b = 9.5 - 14.5$ cm) and mandarin-type fruit ($2b = 5.0 - 7.45$) oblate spheroids. Surface area measurements are useful in considering fruit coatings and waxes and in heat transfer in heating and cooling processes. They are also used in calculating the total quantity of peel oil in the various fruit varieties (Hendrickson et al., 1969; Kesterson and Braddock, 1976). Radii, diameters, and volume determination formulas are used to cal-

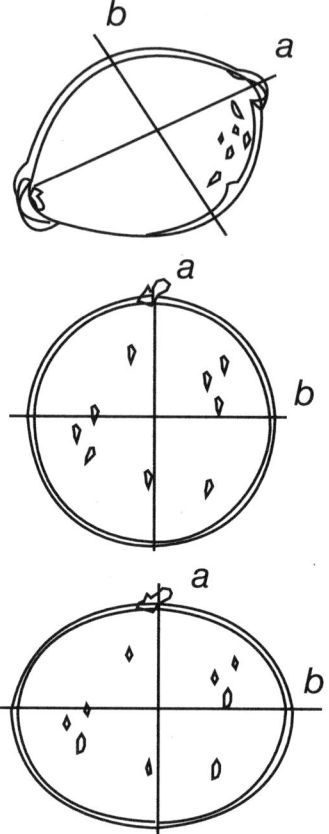

Figure 3.1 Fruit shape and mensuration formulas for lemons, limes, oranges, and grapefruits. V, volume; S, surface area; a, longitudinal radius; b, equatorial radius; e, eccentricity. Lemon and lime (prolate/oblong) $a > b$

$$V = 4/3\pi ab^2$$

$$S = 2\pi b^2 + 2\pi \frac{ab}{e} \sin^{-1}(e) \qquad e = \left[1 - \left(\frac{b}{a}\right)^2 \right]^{1/2}.$$

Orange (round) $a = b$

$$V = 4/3\pi a^3$$
$$S = 4\pi a^2.$$

Grapefruit and Mandarin (oblate/flat) $a < b$

$$V = 4/3\pi a^2 b$$

$$S = 2\pi a^2 + \pi \frac{b^2}{e} \ln \frac{1 + e}{1 - e}.$$

culate volumes for fruit storage, handling and shipping, in machine sorting and grading, and calculating fruit densities (Szczesniak, 1983).

3.1.2 Fruit Components

Following are the various parts of citrus fruit, which can be separately recovered and utilized for by-products. The flavedo is the outer colored portion of the peel. Properties include a wax coating that functions to protect the fruit in its natural environment. The flavedo contains the oil glands that serve as metabolic repositories for the terpenes and aromatic oils essential for the characteristic aroma and flavor of each type of fruit. Beneath the flavedo is the albedo, the white part of the peel. Texture of the albedo is more spongelike than the flavedo, allowing absorption of water and oil. Some of the oil glands may extend into the albedo. The albedo portion of fruit is the richest source of pectin, other carbohydrate polymers, and flavanone glycosides. The interior segments are surrounded by a carpellary membrane and contain the juice vesicles, or sacs. Ruptured juice vesicles after extraction and finishing are referred to as juice pulp. The juice sacs contain the juice and its soluble components, some cellular organelles, enzymes, and essential oils. The membrane and core parts, when contained in the residue from commercial juice extraction, are collectively referred to in the processing industry as the rag. The final component part of the fruit is the seed, with the common oilseed properties of lipids, protein, and carbohydrates. Even so-called seedless citrus fruits contain a few seeds.

3.2 COMPOSITION

3.2.1 Yield and Dry Matter

The dry matter and mass yields of the various fruit components for common fruit varieties are tabulated in Tables 3.1–3.4. As there are industrial machines and processes for separation and recovery of these components, the significance of these data to commercial fruit processing are revealed by closer analyses. First, as fruit matures, the dry solids of the whole fruit and each component increases significantly. At the same time, the juice yields decrease dramatically. This is especially obvious for Dancy tangerines, which have a very high solids content (Table 3.3). Most interest is in the juice solids fraction being as high as possible. In commercial practice, this is not always the best situation because juice quality and yield must be considered. For example, Hamlin orange juice yield is highest in November, but the total juice solids (yield times the percent of dry solids) is no different than for December maturity fruit (Table 3.1). It is important that Valencia oranges at their peak juice yield (Table 3.2) have considerably more juice and higher total solids than Hamlins. Also, with the exception of grapefruit (Table 3.4), considering yield and percent solids, total

TABLE 3.1 Distribution and Dry Solids of Florida Hamlin Orange Components[a]

	Component										
	(% wet wt)					(% dry solids)					
Date	Juice	Albedo	Flavedo	Rag & Pulp	Seeds	Juice	Albedo	Flavedo	Rag & Pulp	Seeds	Whole Fruit
Sept.	40.5	15.9	10.9	31.2	1.5	8.6	21.6	23.5	14.3	30.0	15.0
Oct.	45.8	17.5	10.6	24.5	1.6	8.2	16.9	21.3	14.4	30.0	13.0
Nov.	51.2	13.8	8.6	25.5	0.9	8.9	18.5	24.4	14.4	32.0	13.2
Dec.	47.5	14.5	9.7	27.4	0.9	9.6	19.4	24.9	14.2	35.0	14.0
Jan.	44.3	18.3	10.0	26.6	0.8	10.1	18.0	23.5	15.0	40.0	14.4
Feb.	44.2	18.9	10.3	25.6	1.0	10.7	20.5	25.3	15.6	43.0	15.6
March	41.7	19.5	10.0	28.1	0.7	11.4	20.1	24.0	14.8	42.0	15.5
April	40.1	19.7	8.9	30.8	0.5	11.8	20.7	26.0	14.9	50.0	16.0

[a]Adapted from Hendrickson and Kesterson (1954).

TABLE 3.2 Distribution and Dry Solids of Florida Valencia Orange Components[a]

| | | | Component | | | | | | | | |
| | | (% wet wt) | | | | (% dry solids) | | | | | |
Date	Juice	Albedo	Flavedo	Rag & Pulp	Seeds	Juice	Albedo	Flavedo	Rag & Pulp	Seeds	Whole Fruit
Oct	46.4	16.4	10.8	25.2	1.2	7.8	23.2	22.8	16.9	—	14.5
Nov.	48.7	12.6	10.0	27.5	1.2	8.2	23.3	25.1	16.4	—	14.4
Dec.	48.8	13.6	9.5	27.3	0.8	8.9	25.6	24.7	16.4	—	15.0
Jan.	49.2	13.5	9.3	27.1	0.9	9.9	25.2	26.0	16.4	—	15.5
Feb.	52.3	12.5	9.6	24.9	0.7	9.6	25.9	26.5	16.8	—	15.3
March	51.6	12.4	9.0	26.0	1.0	10.5	21.6	27.0	16.9	—	15.4
April	50.3	13.0	9.3	26.7	0.7	10.8	24.7	25.0	17.0	—	15.9
May	55.0	10.8	10.3	23.1	0.8	11.3	26.7	27.4	18.4	—	16.5

[a]Adapted from Hendrickson and Kesterson (1954).

TABLE 3.3   Distribution and Dry Solids of Florida Dancy Tangerine Components

	Component											
	(% wet wt)						(% dry solids)					
Date	Juice	Albedo	Flavedo	Rag & Pulp	Seeds		Juice	Albedo	Flavedo	Rag & Pulp	Seeds	Whole Fruit
Sept.	37.2	—	16.7	40.7	5.4		8.8	—	29.4	19.6	30.0	17.8
Oct.	45.4	—	16.5	33.9	4.2		8.3	—	27.9	17.2	30.0	15.5
Nov.	43.8	—	14.2	38.8	3.2		8.9	—	25.6	15.0	28.0	14.3
Dec.	41.8	—	11.7	44.5	2.0		9.7	—	25.7	16.8	33.0	15.2
Jan.	28.8	—	10.8	58.8	1.6		10.1	—	23.2	15.2	33.0	14.9
Feb.	34.0	—	10.1	54.1	1.8		11.9	—	27.8	18.4	41.0	17.5
March	27.2	—	6.9	63.5	2.4		13.5	—	30.1	21.2	37.0	20.1

Adapted from Hendrickson and Kesterson (1954).

TABLE 3.4 Distribution and Dry Solids of Florida Marsh Grapefruit Components

	Component										
	(% wet wt)					(% dry solids)					
Date	Juice	Albedo	Flavedo	Rag & Pulp	Seeds	Juice	Albedo	Flavedo	Rag & Pulp	Seeds	Whole Fruit
Sept.	29.5	21.1	8.1	40.5	0.8	8.4	18.0	22.2	13.5	30.0	13.7
Oct.	40.9	22.2	8.8	27.5	0.6	8.5	15.7	19.4	14.2	30.0	12.7
Nov.	46.0	22.0	8.1	23.4	0.5	9.0	16.4	20.1	15.1	32.0	13.0
Dec.	44.4	20.2	8.1	27.0	0.3	9.1	16.2	20.7	13.6	45.0	12.8
Jan.	44.3	21.9	8.2	25.3	0.3	8.7	15.8	19.8	12.9	38.0	12.2
Feb.	50.6	21.2	8.3	19.6	0.3	8.8	16.6	20.3	14.4	44.0	12.6
March	44.2	21.6	7.9	25.9	0.4	8.9	16.7	20.4	12.6	40.0	12.6
April	46.7	21.4	7.6	24.0	0.3	8.6	14.9	19.0	12.3	55.0	11.8

Adapted from Kesterson and Hendrickson (1953).

solids of the rag and pulp fraction is next to that of the juice component. Compared with juice, the percent solids of the flavedo and albedo components are also significantly higher.

3.2.2 Chemical Composition

Detailed compositional data of the edible portions and peel of raw and processed fruit and juices are available (USDA, 1982). The general chemical components present in the fruit parts are stated in the following order. From a processing perspective, citrus fruits are mostly (>85%) water. While this may not be a popular issue for those marketing juice to the consumer, nevertheless it is a fact. This large quantity of water adds cost to processing of the components, since energy is needed for pasteurization or removal during juice evaporation or peel drying. The 15% dry solids fraction of the whole fruit is composed of soluble sugars (10%), fiber (2%), organic acids (1%), amino acids and protein (1%), minerals (0.7%), and oils and lipids (0.3%).

Soluble sugars are the primary constituents of peel, pulp, and rag dry solids. In these fractions, glucose and fructose are about equal to the sucrose content. Besides these major sugars, the peel also contains smaller amounts of xylose and rhamnose (Ting and Deszyck, 1961). Insoluble carbohydrate fibers are pectin, hemicellulose, cellulose, and some lignin. Organic acids in the residue are citric, malic, malonic, and oxalic, with malic and oxalic acid dominating in the peel fraction (Sinclair and Eny, 1947). The pulp, peel, and rag tissues are also the main sources of the flavonoids—naringin in grapefruit, and hesperidin in oranges and most other varieties. The peel is also the source of the major portion of the fruit's essential oils. The primary chemical constituent of these oils is the valued terpene, d-limonene.

As mentioned, the seeds are composed of fats (triglycerides, phospholipids, fatty acids), carbohydrates, protein, fibers, lignin, and minerals. The fats are typically unsaturated, with notable amounts of linoleic and linolenic acids. An interesting feature of citrus seeds is the presence of quantities (1% of seed dry weight) of limonin, which has an intensely bitter flavor.

The juices of the various fruit are like the tissue in that the soluble sugars, glucose, fructose, and sucrose predominate. Total sugars of orange juice are approximately 3% glucose, 3% fructose, and 4% sucrose. Orange and mandarin juices have <1%, grapefruit >1%, and lemon and lime juices >2–3% titratable acids. Because of acid inversion, lemons and limes contain less sucrose than oranges, grapefruits, and mandarins. Citric acid is the main acid in the juices, and in acid fruits may be equal to or greater than the soluble sugar content. Citrus juices contain free amino acids, and the composition has been reported to be of use in detecting adulteration or juice purity. The primary amino acids found in orange and grapefruit juices are proline, aspartic acid, and asparagine (RSK, 1987).

The mineral content of citrus juices is complex and virtually every soluble mineral in the soil absorbed by the tree may be found in the juice. This knowledge has been the basis for use of multivariate statistical testing to determine the geographic origin of citrus juices (Nikdel et al., 1988). In orange juice, potassium (970–2400 mg/L) is in highest concentration, followed by phosphorus (70–240 mg/L), calcium (40–200 mg/L), and magnesium (85–175 mg/L). Vitamin C is the most significant vitamin in citrus juices with a concentration between 25 and 50 mg/100 mL (>100% of daily value in a 240-mL serving) of juice, depending on the product. Orange juice also contains smaller amounts of other vitamins, which have dietary significance, for example, folic acid and thiamin (10% of daily value).

Other constituents of commercial juice depend to some extent on the insoluble suspended matter, referred to as the cloud. These cellular particles cause juice turbidity and contain much of the juice's content of enzymes, protein, flavonoids, and volatile essential oils. Orange, mandarin, and lemon juices contain the tasteless flavonoid glycoside, hesperidin (500–1000 mg/L); grapefruit contains bitter naringin (600–1500 mg/L). The enzyme pectinesterase associated with the tissue has a heat-stable component that must be thermally inactivated at temperatures near 90°C to prevent the cloud from separating from the juice serum. Presence of pectinesterase is significant because the heat necessary for inactivation results in flavor damage to the juice. Water-soluble and tissue pectin concentration of 300–500 mg/L have significance to the mouthfeel and viscosity of juice. The oils and aroma components in the juice are partly bound to the suspended matter and partly soluble in the aqueous phase. Quantities of these components get in the juice from the peel during the extraction process and are important to overall juice flavor. Oil concentrations greater than 0.025–0.03% are less desirable, resulting in a burning sensation in the mouth.

3.3 EVALUATION FOR PROCESSING

3.3.1 Soluble Solids and Acid Content

The combined sensory properties of sweetness and tartness are most important to the fruit and juice quality of citrus. The sugars and acids are the main contributors to the soluble solids. Soluble solids content of the juice is determined based on specific gravity or refractive index of sucrose solutions and is measured on a scale of °Brix. Acids are determined by titratable acidity and expressed in grams per liter or as percent citric (or other) acid. The °Brix/acid ratio is an important sensory indicator useful in defining flavor attributes that depend on maturity and fruit ripeness. Detailed discussions of these methods and measurement techniques can be found in other texts (Redd et al., 1986; Kimball, 1991). Processors must consider fruit maturity in order to meet juice yield and quality standards. As fruit mature during a season, juice soluble solids increase and acid content decreases. These relationships are charted in Figures 3.2 and 3.3, for Florida oranges and grapefruits, respectively. The obvious effect of these changes is that the °Brix/acid ratio increases very rapidly during later

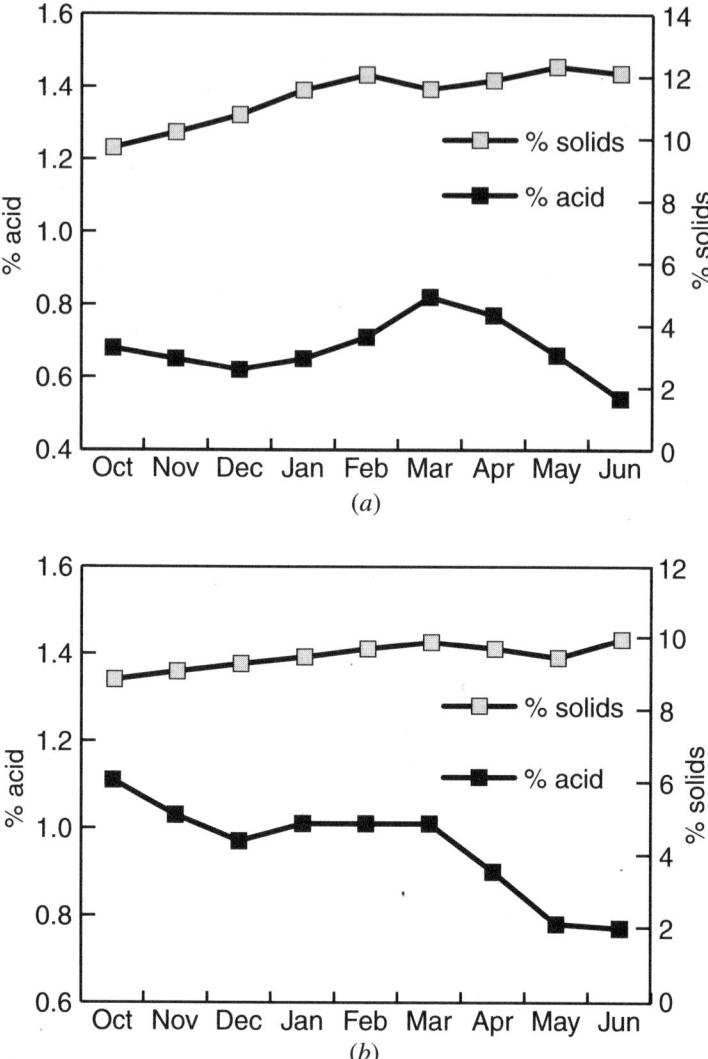

Figure 3.2 Changes in soluble solids and acid during the growing season for (*a*) oranges and (*b*) grapefruit.

months of maturity, dramatically increasing the sensation of juice sweetness over tartness. The more intense flavor of Valencia oranges is reflected in the acid increase during March, when Valencia fruit processing begins (Fig. 3.2). These changes are metabolic in the biochemistry of the fruit, although climatic conditions may play a small role in differences from one season to the next.

Seasonal data of soluble solids and acid content over many seasons for Florida fruit also indicate some trends that relate to juice quality. Orange juice

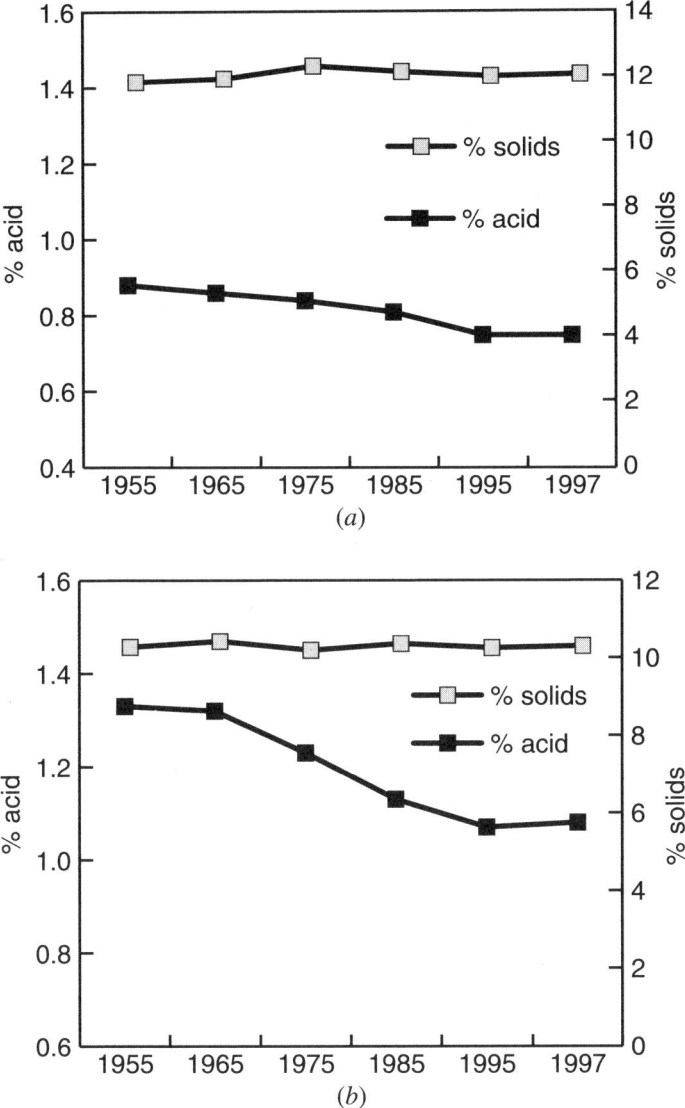

Figure 3.3   Changes in soluble solids and acid over many seasons for (*a*) oranges and (*b*) grapefruit.

average soluble solids have increased slightly, while a more noticeable decrease in acidity has occurred from the 1950s to the present (Fig. 3.3). This result is expressed in an increase in the average °Brix/acid ratio. The reasons for the significant decrease in acid are related to severe freezes in Florida during the 1980s, which resulted in many newer plantings below the freeze zone in South Florida. The large amount of fruit from young trees, climatic differences (more sunlight, warmer days and nights, more rain), and some varietal planting shifts to early and midseason fruit affected the quality of the juice. A similar scenario is observed for the °Brix/acid ratio of grapefruit juice (Fig. 3.3). Besides South Florida plantings, a factor in grapefruit juice characteristics has been the large shift to seedless (both white and colored) fruit, which tend to have lower solids and acid and less flavorful juice than seedy varieties.

3.3.2 Sensory Evaluation

A necessary requirement for processing fruit is that the juice has desirable sensory attributes. Sensory testing of juice flavor, color, and other characteristics is widely practiced where juice is manufactured and packaged. Because of stringent governmental requirements and standards, Florida has very organized official taste panel methods for juice products. The Florida Citrus Processors' Association contracts one of the official taste panels with the USDA. A unique feature is that this panel is a consumer preference panel composed of local citizens and processing plant personnel. The panel tests frozen concentrated orange juice (FCOJ) weekly on a nine-point hedonic scale rating from "extremely poor to excellent." Copies of the test results and coded juice rankings are sent to participating processors with only their product identified.

Another weekly flavor panel evaluates FCOJ, concentrated orange juice for manufacturing (COJM), pasteurized and chilled orange juice (POJ, COJ), single-strength orange juice from concentrate, and canned orange juice. This panel at the USDA Processed Products Office in Winter Haven, Florida, defines the described products according to the Standards of Identity for USDA Grades by a point rating system. For example, a grade A orange juice would have a "very good flavor," "fine, distinct and similar to fresh juice" with a 36- to 40-point rating. There is also USDA in-plant grading by official inspectors. These inspectors taste and evaluate all of the 100% juices that have USDA Standards of Identity. The evaluation includes tests for °Brix and acid and examination for defects such as mold mycelia, embryonic seeds, particles of evaporator scale, and other extraneous matter.

The citrus industry outside of Florida is mostly dependent on in-house quality assurance standards and sensory tests. Factors for selecting taste panelists for routine juice product evaluations need not be too complicated. In general, it is advisable to use 5–10 trained panelists and tell them the type of sample (OJ, FCOJ, POJ, etc.) they are evaluating, but not much else. If they are told the samples are from high-quality fruit, the rating will reflect that, even though it may not be true. Random sample coding and presentation to the panelists is also very important to prevent bias. The number of samples at a taste session should also be few (no more than 2–4 for citrus juices). Testing conditions, such as the panel room lighting and temperature, sample temperatures, time of day, and procedure are all important to the results. Finally, the sample questionnaire and the type of taste test should be as simple as possible. Common tests for citrus juices are the triangle test, paired comparison, and product ranking.

REFERENCES

Hendrickson, R. and Kesterson, J. W. (1954). Hesperidin, the principal glucoside of oranges. Agr. Exp. Sta. Bull. No. 545. University of Florida, Gainesville, FL, 43 pp.

Hendrickson, R., Kesterson, J. W., and Ting, S. V. (1969). Peel oil content of Valencia oranges, *Proc Fla. State Hort. Soc.* **82**:192–196.

Kefford, J. F. and Chandler, B. V. (1970). *The Chemical Constituents of Citrus Fruits*, Adv. Food Research, Suppl. 2, Academic, New York.

Kesterson, J. W. and Braddock, R. J. (1976). Total peel oil content of the major Florida citrus cultivars, *J. Food Sci.* **40**:931–933.

Kesterson, J. W. and Hendrickson, R. (1953). Naringin, a bitter principle of grapefruit. Agr. Exp. Sta. Bull. No. 511. University of Florida, Gainesville, FL, 35 pp.

Kimball, D. A. (1991). *Citrus Processing Quality Control and Technology*, Van Nostrand Reinhold, New York.

Nagy, S., Shaw, P. E., and Veldhuis, M. K. (1977). *Citrus Science and Technology*, Vols. 1 and 2, AVI, Westport, CT.

Nikdel, S., Nagy, S., and Attaway, J. A. (1988) "Trace metals: Defining geographical origin and detecting adulteration of orange juice." In *Adulteration of Fruit Juice Beverages*, S. Nagy, J. A. Attaway, and M. A. Rhodes, Eds., Marcel Dekker, New York, Chap. 5, pp. 81–105.

Ranganna, S., Govindarajan, V. S., and Ramana, K. V. R. (1983). Citrus fruits: Varieties, chemistry, technology, and quality evaluation. Part II. Chemistry, technology, and quaity evaluation. A. Chemistry, *CRC Crit. Rev. Food Sci.* **18**(4):313–386.

Redd, J. B., Hendrix, D. L., and Hendrix, C. M., Jr. (1986). *Quality Control Manual for Citrus Processing Plants*, Vol. I, Intercit, Safety Harbor, FL.

RSK (1987). *RSK-Values. VdF*, Flüssiges Obst Gmbh. D-5429 Schönborn, Germany.

Sinclair, W. B. (1972). "Composition of the edible portions" (Chap. 2); "Composition of the peel, rag and seeds" (Chap. 3). In *The Grapefruit*, University of California Press, Riverside, CA, pp. 62–280.

Sinclair, W. B. (1984). *The Biochemistry and Physiology of the Lemon and Other Citrus Fruits*, University of California Press, Oakland, CA.

Sinclair, W. B. and Eny, D. M. (1947). Ether-soluble organic acids and buffer properties of citrus peels, *Bot. Gaz.* **108**:398–407.

Szczesniak, A. S. (1983). "Physical properties of foods: What they are and their relation to other food properties." In *Physical Properties of Foods*, M. Peleg and E. B. Bagley, Eds., Avi Publishing, Westport, CT, Chap. 1, pp.1–41.

Ting, S. V. and Deszyck, E. J. (1961). The carbohydrates in the peel of oranges and grapefruit, *J. Food Sci.* **26**:146–152.

USDA. (1982). *Composition of Foods. Fruits and fruit juices. Oranges*. Agricultural Handbook No. 8–9, Washington, D.C.

CHAPTER 4

JUICE PROCESSING OPERATIONS

Juice is the primary processed citrus product, amounting to greater than one-half of the fruit mass. Technology for juice recovery must be reliable, efficient, versatile, and sanitary. Also, the choice of juice recovery method has important consequences for recovery techniques applied to some of the major by-products. The following discussion will describe some of the important variables relating to extraction methods and other process operations required in commercial juice recovery.

4.1 EXTRACTION

After final fruit sanitizing and grading, the sizing operation delivers fruit to the juice extraction line as described in Chapter 2. Sizing is very critical to assure juice quality and efficient extraction and determines the placement of juice extractors in a processing line. This placement arranges extractors at the beginning of the line designed to handle smaller fruit, increasing fruit sizes to the end, where extractors designed for the largest fruit are placed. Juice quality and yield per fruit mass is improved for smaller rather than larger fruit, justifying changing extractor variables to compensate for fruit size, besides the inherent efficiency of matching extractor cups to fruit diameter. When considering juice quality, fruit quality is probably most important, followed by variables controlled at the juice extractor. These variables (pressure, clearance, cup size, etc.) are determined mostly by fruit size, juice yield, and quality desired by the processor. Designs of the major commercial citrus juice extractors include those by the FMC FoodTech Citrus Systems Division and Brown International Corporation. Detailed operational descriptions of these machines may be obtained from the companies and have been published (Kimball, 1996). Some Italian-made machines, such as the Indelicato Polycitrus line, are also in use for essential oil and juice recovery.

4.1.1 Extractor Automation

Modern citrus processing plants using the Brown International or FMC juice extraction systems have their juice room processes controlled by computers. The plants desiring to utilize the advantages of automation cooperate with the extractor manufacturer to design a system suitable for their needs. Such systems may have only limited control of automation for a few unit operations, such as the extractors and finishers or they may extend from the juice control room throughout the plant operations. Real-time operation of the entire plant at the juice room control center may range from fruit receiving to the evaporator and blend tanks, including not-from-concentrate (NFC) processes. The systems are set up to allow operators a visual view of the process flow streams on a computer monitor and to interact and override automation at a point, if necessary. The primary advantage of this automation is increased efficiency through optimization of the flow rates of the various streams. In addition, equipment cleaning and sanitation processes are part of this scheme, contributing to maximizing process operations. As for individual automated unit operations, the computer logs a permanent record of the process, which is useful for maintenance, quality assurance, and accounting purposes.

4.1.2 Brown International Corporation Extractors

Brown International Corporation (Covina, CA) (the Florida division is named Automatic Machinery & Electronics, Inc., Winter Haven, FL) manufactures a main line of citrus juice extractors that recover juice by reaming. The company also makes machines that recover juice using a squeezing principle. Pictures, diagrams, and detailed operational information of these machines are available from the manufacturer and other publications (Kimball, 1996).

The primary reamer is the Model 700, which has the reamers and cups mounted in a vertical plane. The Model 700 has the capacity to process greater than 700 fruits/min, provided that adequate fruit is delivered to the machine and the reamer cups each receive a fruit half in a continuous sequence. Newer designs of this machine (Model 720) have been smoothed inside to facilitate automatic cleaning and sanitation regimes. Reduced cleaning time allows increased operation time. Figure 4.1 is a diagram of the functional components of this extractor. The reamer heads are attached to turning shafts on the two faces of a revolving cylindrical drum. A feed wheel transfers fruit past a knife, which halves each fruit, to a chain of cups. The cups deliver the fruit halves to the reamers, the juice is collected beneath the reamers and the reamed half-peels are ejected from the cups by a device called a kicker, which fits into a

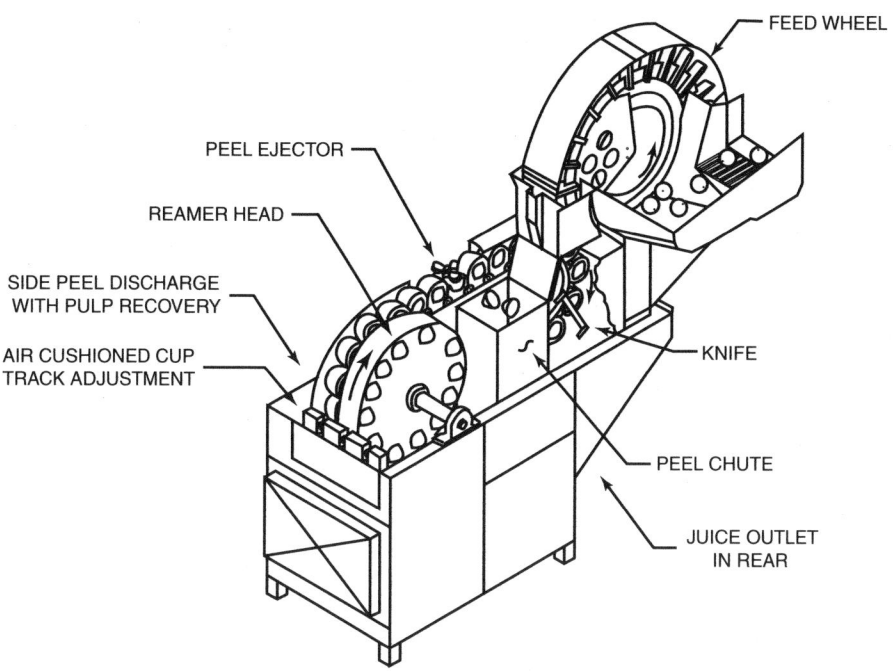

Figure 4.1 Working components of a Brown Model 700 juice extractor. (By permission of Automatic Machinery & Electronics, Inc., Winter Haven, FL).

slot in the cups. Operational parameters of the Model 700 are illustrated in Figure 4.2. Fruit halves fit into the cup and adjustable air pressure holds the cup against the spinning reamer, releasing the juice. Working in conjunction with the air pressure control, another parameter is a mechanical setting to compensate for different peel thickness encountered during processing of a run of fruit. These parameters prevent the reamer from overextracting down into the albedo layer of fruit with thicker peels, yet allow complete juice recovery from fruit with thinner peels.

There are two product streams recovered at the juice extractor, the pulpy juice and the peel residue. These streams are recovered beneath the extractor lines for further processing. The extracted juice stream goes to a primary finisher with a larger-hole screen size (0.125–0.188 inch) to separate the pulpy juice from the membranes, rag, and seeds. The pulpy juice stream goes to another finisher with a smaller screen (0.020 inch) to effect separation of the juice and pulp. The peel residue and primary finisher discharge may be sent to another by-product process or to the feed mill for cattle feed manufacture.

The juice quality is primarily dependent on the fruit quality; however, the extractor air pressure and peel clearance settings allow some control over quality. Higher pressures result in more peel oil in the juice, while lower pressures give less. Juice yield is directly related to the peel thickness and extractor

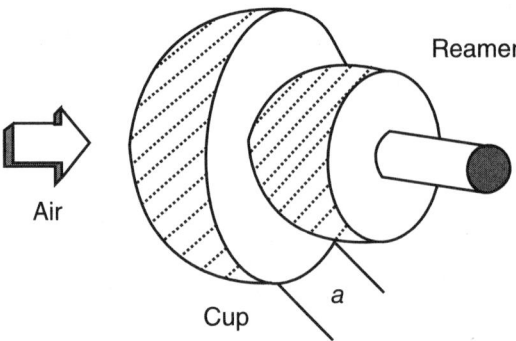

Figure 4.2 Operational diagram of a reamer-type citrus-juice extractor indicating application of air pressure, the fruit holding cup, the reamer, and the peel clearance setting (designated by *a*).

settings. For example, if an extractor is set for thick peel and the fruit has thin peel, juice yield will be less. If the peel is thick and there is a thin setting, the reamer would be too tight (into the albedo) for good juice quality. Thus the air pressure helps maintain constant yield. A feature of reamer-type extractors is that some visual evaluation of the performance is possible by trained technicians or plant personnel. Simply, examination of the reamed extracted fruit halves will allow one to see the degree of extraction. Best juice yield and quality occurs when some membrane is left in the half-peel and reaming into the albedo does not happen.

4.1.3   Other Brown Extractors

Brown Model 400 extractors are also reamers. The 400 designation, in general, indicates that these extractors can process up to 400 fruits/min. These machines have the reamers mounted horizontally, rather than vertically, and operate at slower speed. In the United States, the Model 400 is mostly used for grapefruits and large-sized oranges. In orange juice installations where Model 700 extractors are used, the Model 400 machines are placed at the end of the lines to take the large fruit from the sizers, allowing more efficient juice recovery. The Model 400 extractor has no peel thickness compensation.

Model 1100 extractors (1100 fruits/min) are vertically mounted extractors that operate on the principle of squeezing (instead of reaming) the cut fruit halves against perforated grids separating revolving disks. These fast machines are used to recover juice from tangerines, tangors, and small oranges. Juice from these extractors has a high oil content due to contact with the peel during recovery. Sometimes baffles are put into the juice collection zone to allow separation of juice squeezed from the front and back regions of the disks. In this way, lower and higher oil content juice may be recovered separately, allowing some control over juice quality.

Brown also makes an extractor (the Pine-O-Mat) for recovering pineapple juice. The Pine-O-Mat is mentioned here because its precursor, the Citromat, was used to process oranges and grapefruits in the early years of citrus juice extractors. This extractor has the capacity to process 9 mt (10 tons) of pine-apples/hr, yielding 4.5 mt (5 tons) of juice/hr. Operation involves a feeder where pineapples are fed longitudinally to saws that slice the fruit into halves. The halves are oriented with the cut surface outside, onto a revolving drum fixed with spikes, pressing the fruit against a perforated grid. As the drum rotates, the clearance between the grid and drum decreases, squeezing the juice through the grid and ejecting the peel from the back of the extractor. The juice quality is good because there is minimal contact of the juice with the rind and its oils, compared with pineapple juice presses. This is not the case when the Pine-O-Mat is used for citrus juice extraction, because of the ease of rupturing the peel oil glands, allowing oil to enter the juice.

4.1.4   FMC Corporation Extractors

All of the FMC extractors operate on the principle of instantaneous separation by squeezing the juice from the fruit. The major differences between extractors for the various fruit types are the number and size of the fruit cups and the speed of the extraction cycles. Detailed information about the various FMC juice extractors is available from the company (FMC Food Processing Division, Lakeland, FL). FMC also maintains a home page on the Internet (FMC, 1998). The design allows shape and size sensitivity, stability of settings, and flexible configuration for various processing environments and allows operational parameters to be changed to allow variability in juice yield and quality. For example, machines used for grapefruit processing have larger cups, while tangerine machines have smaller cups. The most common orange extractor has five cups and an extraction cycle of 100 strokes/min. As for other extractors under commercial conditions, the reality of not being able to have all five cups filled with fruit on each and every stroke means the actual extraction rate is less than the theoretical value of 500 oranges/min.

In a processing plant, the extractors are arranged in linear rows, with extractors at the beginning of the line set up to handle smaller fruit and those at the end to handle the largest fruit. Fruit missed by the extractors is carried back to the surge bin by a conveyor belt as previously diagrammed (Fig. 2.1). All extractors are arranged for easy access and can be set to accommodate automatic cleaning and sanitation. A simplified description of the operation of the FMC extractor is illustrated in Figure 4.3, parts 1, 2, and 3.

FMC extractors generate a number of product streams at the extractor during the juice extraction cycle. Because these streams become the feedstock for by-products manufacture, they are mentioned here. By virtue of a strainer tube attached under the lower extractor cup, separation of the pulpy juice (the primary product stream) occurs. The strainer tube serves as a prefinisher. The plugs of peel cut from the fruit by the upper and lower cutters, seeds, and

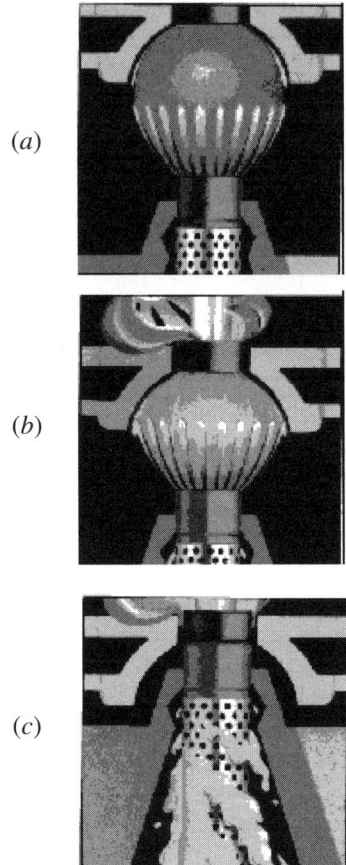

(a)

(b)

(c)

Figure 4.3 Steps in the juicing process of an FMC juice extractor. (a) Upper and lower cups start to converge and the upper and lower cutters cut two holes in the fruit. (b) Upper and lower cups continue converging and the peel is separated from the fruit. (c) Peeled fruit is squeezed into the strainer tube, instantaneously separating juice from seeds and the rest of the fruit. (By permission of FMC FoodTech, Citrus Systems Division, Lakeland, FL.)

membrane are separated by ejection from the bottom of the tube, while the juice and ruptured juice vesicles pass through the 0.040-inch-diameter perforations (Fig. 4.3). The ejected material and the outer peel residue from the fruit falls into a screw conveyor beneath the extractor line for transport to the feed mill. In some plants, extractor lines are set up with a small conveyor to recover the ejected material from the stainer tube separately from the peel residue. This fruit core and membrane material can be washed with water to recover soluble solids, core wash, as a by-product. The pulpy juice stream flows to a finisher, which separates the juice pulp from the juice. The pulp may be considered one of the by-products. If necessary, juice is pumped from the finisher to a centrifuge for removal of defects and to adjust the pulp content or viscosity before further processing as pasteurized juice, concentrate, or other specialized product.

As the upper and lower cups converge and the juice passes down the strainer tube, the peels (flavedo and albedo) are forced out through a small circular opening and the fingers at the top of the upper cup. Oil is released by compressing the small glands in the flavedo, which contain the peel oil. A water spray fixed in a ring to the upper cup washes the flavedo as the peel is forced through this opening, forming an oil/water emulsion with the released peel oil. The force of this process in the extractor cup breaks loose some small pieces of peel, which are referred to as peel frits. The frits and the oil emulsion flow over the back manifold of the extractor, pass through a finisher to remove the frits, and on to the peel oil recovery process.

4.1.5 Other FMC Extractors

The FMC Premium Juice Extractor is a machine option similar to the primary commercial machines in use in most citrus plants. The major difference between other FMC extractors and the premium juice extractor is that the machines are set up for recovery of high-quality NFC juice. Not-from-concentrate juice recovery requires attention to some details of quality, which are not as important when manufacturing juice for concentrate (oil control). Higher quality implies better flavor and stricter monitoring of the standard quality control attributes of juice. Because of this, there is less margin of error in many of the quality parameters related to the juice recovery process. The first of these is the fruit itself. Only high-quality fruits are used in a premium juice extractor line, and the extractors are set up for only two fruit sizes. The setup also results in less peel oil in the juice, which improves the flavor. A lower extraction pressure and decreasing the plug cutter diameter from 1 to 0.75 inch also improves the juice flavor, but may result in approximately a 2% juice yield loss.

FMC also makes a small extractor, the Fresh 'n Squeeze™, for local fruit vendors and small, fresh, unpasteurized juice markets. The machine only has one fruit cup and uses a similar operating principle as the large commercial extractors. This machine can extract juice from the various citrus varieties at a rate of approximately 20 fruits/min., has a 40-lb fruit capacity, and a juice reservoir capacity of about 3.5 gal.

4.1.6 Italian and Other Extractors

A common commercial machine made in Italy is the Polycitrus™ extractor [Fratelli Indelicato, Giarre (Catania), Italy]. This extractor is used in combination with an oil recovery machine, which is mounted above the juice press. The peel oil extraction occurs by water sprays in a closed, cylindrical drum environment as the fruit flavedo is scarified, or rasped, on rotating cylinders of sharp, punch-perforated stainless steel. The rasped fruit falls to the extractor where the juice is recovered from cut halves by pressing against perforated stainless steel sheets. Unless care is taken in the oil machine, the juice quality may be poor due to its high peel oil content and other extracted peel constituents. Because of juice contact with fruit surfaces in the press, sanitation of the fruit, oil extractor, and juice press equipment must be watched very closely, otherwise high microbial counts may occur in the juice stream. These machines have a high fruit throughput. One model is capable of processing approximately 15–20 mt of oranges/hr. Although it contributes to inefficiency, another feature of this type of extractor is that fruit can be processed without sizing, which may have some advantages in small to medium-sized, minimum investment operations.

A small, point-of-sale extractor similar to the FMC Fresh 'n Squeeze extractor, uses a patented principle to peel the fruit before juicing (Mendes, 1997). Claims for this machine are that the juice has a low peel oil content and the ability to extract 180 fruits/hr to yield >30 gal/hr (8 L/hr) of juice.

4.2   FINISHING

Finishing is the term applied to the physical separation of portions of the pulp and other fibrous material from the main juice or by-product stream from the extractors. In fruit juice processing operations, finishing is performed by machines with either a rotating screw or paddles, surrounded by a cylindrical, fixed, porous, stainless steel screen. Product liquid passing through the screen is recovered in a juice pan mounted under the screen, and the solid material is discharged at the end of the finisher distant from the entry stream. For some applications, the discharge might be considered the product stream. The two common types of finishers are referred to as either screw or paddle finishers and have different applications in citrus juice and by-product processing technology.

4.2.1   Screw Finishers

Screw-type finishers are most commonly used for juice pulp reduction applications and in peel oil separations to remove larger particles from the oil emulsion stream prior to centrifugation. This finisher operates by the turning screw working to concentrate the pulp (pomace) against a pressurized discharge endplate, or cone opening, forcing the juice out the holes in the screen, draining into the collection pan. Some important design feature performance variables affecting screw finisher operation are considered in the following discussion.

The screw design may be amenable to a number of configurations, such as the number or closeness angle of the screw flights, the diameter and shape (taper) of the screw, or the clearance between the screw edges and the screen (Fig. 4.4). Generally, there is a close tolerance between the screw and the screen. Some screw designs have conical tapers from either the juice feed input or discharge ends. Since rotation of the screw carries the juice and pulp down the screen, revolutions per minute (rpm) is another important design variable to aid in squeezing and mechanically drying the pad of pulp formed at the discharge end. Lower rpm will decrease the throughput rate, and more work will result in drier pomace, more suspended particles forced through the screen, and higher viscosity juice. These variables can be studied to maximize finisher capacity or throughput and juice quality.

The most common design variables related to the screen are the diameters of the holes through which the juice or liquid passes. Common juice finisher

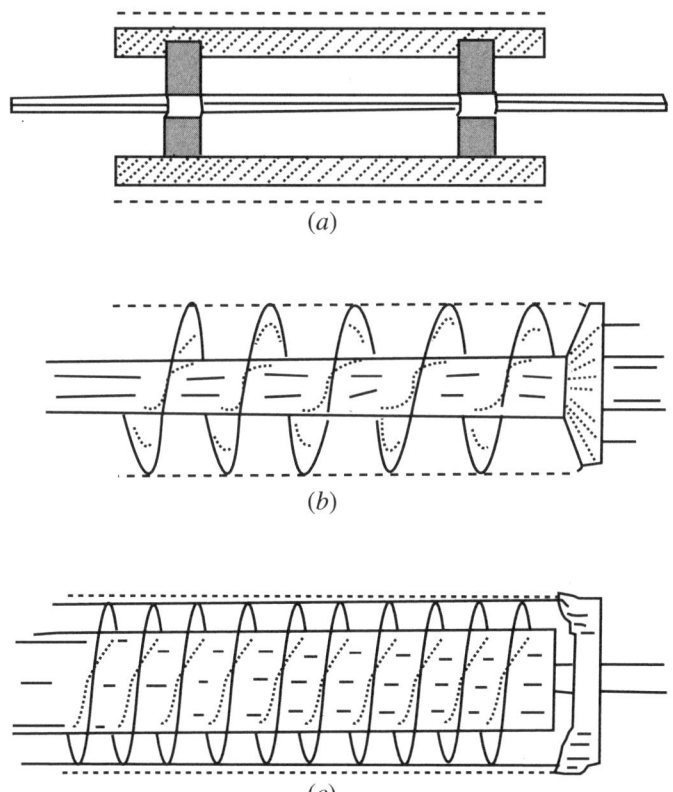

(a)

(b)

(c)

Figure 4.4 Sketches of the working components of (a) a paddle, (b) a single-flight, and (c) a double-flight screw finisher. (By permission of Automatic Machinery & Electronics, Inc., Winter Haven, FL.)

screen sizes depend on the extraction system (Brown or FMC) and application, as described above. Both systems require two finishing steps, the first with larger holes (e.g., 0.125 inch) and the second with smaller holes (0.020 inch) for the pulpy juice stream from the first step. FMC systems can use 0.025 inch tubes and no finisher or 0.040 inch tubes and 0.020 inch screens. As stated earlier, the first finishing in the FMC process occurs in the extractor strainer tube, while the Brown process requires two finishers. Screens may also be designed with different hole sizes (or shapes) on the surface of the same screen. Depending on the properties of the pulp, the smaller hole sizes should reduce insoluble material and provide juice with lower pulp content and smaller particles. A finisher with a large screen hole size (0.125 inch) is termed a scalper. This finisher is used to remove membranes and seeds or for pulp recovery. For concentration and packaging of pulp, small hole screens and close tolerances are used. Small hole screens are also used for pulp washing after the primary finisher.

The discharge end of a screw finisher may be designed for pressure or other control, by affecting the clearance of the pulp discharge orifice. Higher pressures or less clearance may force more pulpy material out through the screen into the juice stream. This may increase juice yield with drier pulp but result in juice with increased viscosity and insoluble solids. Lower pressures, or more clearance, decrease yield and may make juice more like a beverage (less consistency). Lower pressure may improve flavor by reducing bitterness from extracted juice pulp components. Also important to these variables is the pulpy juice stream feed rate. Depending on the capacity, or volume, of the finisher, a lower feed rate will result in more work done by the screw (longer residence time), drier pulp, and higher juice yield. Higher feed rates yield less juice and correspondingly wetter pulp.

Another option available in the finishing operation allows performance variability specifically related to juice quality and offers some options for by-products recovery. It is possible during juice recovery to put a baffle(s) in the juice collection pan, allowing separation of the juice from the different zones along the finisher screen. An example of this technique follows: One baffle in the pan collects juice from the high-pressure end and from the lighter, free-flowing, no-pressure entry zones of the finisher. The juice from the heavy end zone will have more viscosity, higher pulp, and oil contents and may have some astringency and off-flavor. This juice might go separately to an evaporator for concentration and essence recovery or another process stream in the plant. The juice from the lighter end zone may have better flavor, lower viscosity, and less oil. This stream (free-run juice) may be sent to a pasteurizer for packaging a higher-quality, single-strength product.

4.2.2 Paddle Finishers

Paddle finishers have rotating paddles inside the screen, as indicated in Figure 4.4. They have no cone or restriction in the discharge end to impede the flow of the discharged pulp or pomace. Design features that affect performance are rpm and angle (or pitch) of the paddles and the clearance between the paddles and the screen. For higher rpm, there will be more centrifugal force, which will result in greater juice yield, more pulp in the juice, and drier pulp out the discharge. At decreased rpm, the result is lower yield, less pulp in the juice, and wetter pulp. The angle or pitch of the paddles is the offset from the plane of the shaft. More offset at a fixed rpm decreases the juice retention time and will increase the throughput or finisher capacity. Less offset or pitch increases the retention time. Industry applications of pitch are the use of higher offsets for products with poor flow characteristics (e.g., peel bits or very pulpy juice). Lower offsets are applied to juice refinishing and peel oil recovery, where a fine screen (0.010 inch) and longer retention times may be necessary to reduce fine pulp. Close tolerance screen/paddle clearance settings (distance between paddles and screen) help clean the screen by a sweeping action of the paddles. A wider tolerance allows formation of a cake between the paddles and the screen, which helps filter the juice through the screen but results in wetter pulp. A common paddle finisher screen size is 0.020 inch.

Citrus applications for paddle finishers include decreasing juice pulp and reduction of peel and membrane particles in peel oil emulsions. The latter process helps increase the efficiency of oil recovery during the centrifugation steps. Other juice applications involve putting baffles in the pan to collect juice from the heavy end zone. Only this fraction will be sent to the juice centrifuge for defect and pulp removal, reducing the size of the centrifuge needed. In general, for this application, approximately one-half of the pulp is in one-fourth of the juice flow. Paddle finishers are also commonly used in pulp recovery processes, where liquid reduction of the pulp stream is required. In this application, the pulp, not the liquid, is the product stream.

4.2.3 Turbo Filter™

A type of juice finisher that has been in use in the dairy industry has been modified and adapted to citrus juice finishing in Brazil and Florida. Termed a turbofilter in the citrus industry, this machine, the Turbo Filter™ (Mecat Service Ltd., Araraquara, Brazil) has a fabric filter sleeve in place of a traditional porous stainless steel screen (Fig. 4.5). The insoluble floating pulp in the pressurized feed stream is filtered through a sleeve of chemically inert 110-μm (130 mesh) fabric with high mechanical strength. Turbulence of the stream is created by rotating blades and minimizes fouling the filter through the sweeping action of the liquid. Claims are better quality juice, with improved flavor and color due to the soft impact against the nonmetal fabric surface. For citrus juice pulp recovery and reduction, use of the Turbo Filter may replace some use of a centrifuge (e.g., in defect removal). Because the discharged fines and pulp are wet, there is some juice yield loss; however, solids may be recovered by pulp-washing in installations using this equipment. Simple maintenance require-

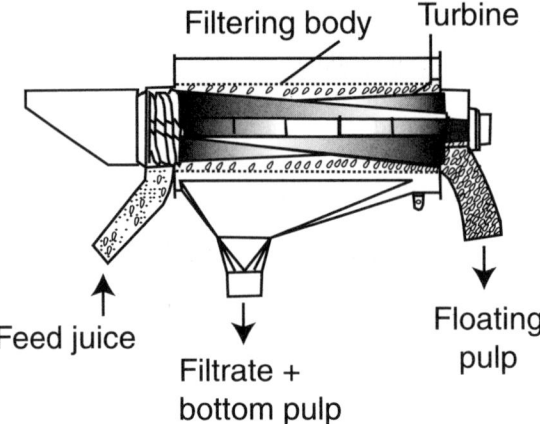

Figure 4.5 Diagram of a Turbo Filter for juice pulp reduction and floating pulp recovery by a continuous, turbulent filtration process. (The Turbo Filter is manufactured by Mecat Filtraçöes Industriasis Ltd., and is patented by Attilio Turchetti, president of this company.)

ments include occasional ball-bearing replacement and changing the retainers and filter sleeves.

4.2.4 Finisher Applications

Pulp reduction of juice is the most common finisher application. Except for juices prepared with extra pulp for consumer preferences, pulp is reduced to affect the quality of specific juice products. For example, juice with lower pulp has lower viscosity and will have more beverage-like sensory properties. This juice will also perform better during evaporator concentration because of the potential of pulp to increase viscosity significantly at higher °Brix (Vitali and Rao, 1984). Use of finishers can lower pulp to about 8% v/v, with an average of 10–14%. For lower values, the juice must be centrifuged, adding to the cost because of additional processing and lost yield. If juice flow rate to the finisher is reduced too low, the finisher will perform too much work, and the result is increased pulp in the juice. Also, mechanical degradation of the pulp will result in higher amounts of pectin in the juice, increasing viscosity. In Florida, standards for FCOJ limit the pulp content to a value of 12% for USDA Grade A. Finishers also are used to balance juice quality and yield with hard-squeeze or soft-squeeze extractor settings. Besides viscosity, finisher performance can affect juice cloud, color, flavor, flavonoid glycosides, and pulp level. There is evidence that finisher settings do not affect the level of oil in the juice, which is a function of the juice extractor.

Finisher performance may be measured by a quality control procedure, termed the Quick Fiber Test (Redd et al., 1986). This test is used to determine

TABLE 4.1   Finisher Settings Related to the Quick Fiber of Pulp from Orange and Grapefruit Juices Recovered from Two Types of Commercial Extractors

	Quick Fiber Value			
	Brown Extractor		FMC Extractor	
	Loose Finish	Tight Finish	Loose Finish	Tight Finish
Orange	110–130	70–90	180–210	130–150
Grapefruit	120–140	90–110	180–210	130–150

the relative dryness of the pulp discharged from the finisher. The method involves mixing 200 g of pulp with 200 g water, placing on a vibrating screen (20 mesh for Brown, 40 mesh for FMC pulp), and measuring the milliliters of liquid recovered. The quick fiber values are affected by the finisher performance and properties of the pulp, such as the type and maturity of fruit. However, general values for finisher conditions are illustrated in Table 4.1. Interpretation of these values follows the logic that the drier pulp will absorb some of the added water and yield lower quick fiber values. Conversely, wetter pulp has higher values. The correlation to juice yield is such that higher quick fiber wet pulp contains more free juice, thus a lower juice yield than dry pulp with lower quick fiber values.

Flavonoid glycosides are components of the fruit and pulp tissue that have a compositional relationship to juice extraction and finisher performance parameters. For grapefruit juice, harsh extraction and hard finishing will extract more naringin from the pulp into the juice and content could range from 700 to 1000 ppm. The juice would have poor flavor quality because naringin is very bitter. If the juice has less than 300 ppm, this could be indicative of too soft a finishing and low juice yield. The same is true of orange juice for hesperidin content. Although this flavonoid is tasteless and poorly soluble, the concentration in the juice is very much different than for naringin. Values of 125–150 ppm in the juice imply poorer quality and hard finishing, while if less than 75 ppm, conditions are too soft and juice yield is low.

4.2.5 Finisher Automation

Screw finishers can be operated in a fixed mode or under automated performance control. Modern citrus processing involves computer control to maximize juice yield and quality. Historically, finisher pressures have been set by monitoring the quick fiber and juice characteristics. Once set, the values were constant unless some condition changed. Due to changes in variables of juice composition or the stream flow rate during operation, real time manual changes are difficult to implement. Thus, automatic control of a finisher has tangible benefits. For automation of a finisher, pressure is an important control point. Citrus applications will determine finisher controls depending on the desired juice characteristics.

4.2.5.1 FMC Automated Finishers

This technique requires use of juice surge tanks and measuring and maintaining a constant juice flow rate to the finisher (FMC, 1993). The application functions by setting the desired quick fiber of the discharged pulp and the interval for the automatic spray ring cleanup cycle on a programmable logic controller (PLC). The pulpy juice is collected in an enclosed surge tank and then pumped at constant flow rate to the finisher(s). Juice line flow meters send data to the PLC, which determines air pressure to the finisher discharge valve for the quick fiber set point. This control system automatically adjusts air pressure when juice flow variations occur to maintain a constant quick fiber. Automated finishers have a piston at the bottom discharge and the screw does not extend beyond the screen. Also, at preset times, the finisher air pressure valve will be relieved, all pulp will discharge, and the automatic high-pressure water spray cleanup cycle starts.

4.2.5.2 Brown Automated Finishers

Automated control of Brown finishers functions by keeping a constant temperature difference (ΔT) between the feed and exit juices (Waters et al., 1987). The juice quality and pomace dryness determine the desired ΔT set point between the incoming feed juice and the discharged juice or pulp. The control system is based on the mechanical heat equivalent resulting from the work performed when the pulp is compressed in the pad zone inside the screen at the finisher discharge. The temperatures are determined by thermistors in the incoming and heavy end zone juice streams or pad area pomace discharge. The computer determines the air pressure necessary to maintain a constant ΔT between the input and discharge juice and collects data at intervals (e.g., each minute) from the finisher. The pomace discharge pressure valve is pneumatically adjusted to compensate for changing operating conditions, such as feed rate. The result is to automatically maintain constant juice quality, yield, and pomace dryness.

4.3 CENTRIFUGATION

Centrifuges in service for citrus processing are primarily solid bowl/sedimentation type of clarifiers used for solid–liquid or liquid–liquid separation applications. Some horizontal decanter-type machines are also in use for dewatering and clarification applications, where continuous suspended solids removal is necessary. Machine uses vary from quality control laboratory to major processing stream, batch, or continuous operations. Centrifuge theory is not the subject of this discussion, but for solid–liquid separations, the sedimentation velocity of a particle depends on the densities of the solid and liquid phases, the particle diameter, and the viscosity of the liquid phase. The lighter particles accelerate inward faster with highest g force at the inner surface of the bowl.

Citrus processing centrifuge applications are mostly clarification/solids recovery where the product is not solids-free (e.g., separation of pulp from juice), clarification where the product is a clear liquid (e.g., peel oil recovery or very low pulp juice), and solids recovery where the product is the solid material (e.g., pulp dewatering). Most centrifuges are automated for operational efficiency and considerable information is available from manufacturers. Good descriptions of citrus processing applications have been published (Bott and Schöttler, 1989; Tetra Pak, 1998).

4.3.1   Juice Pulp Reduction

The most common application of centrifuges in citrus processing is for juice defect removal and pulp and viscosity reduction. Extracted juice from finishers may contain small black particles or other undesirable visible material that is removed by centrifugation, improving the appearance to the consumer. Also, centrifugal control of the pulp level is more exact than by finishers to meet consumers' preference for low pulp, low viscosity juices with between 6 and 12% pulp. There are also consumers who prefer higher pulp-added juice products in the range of 12–24%. Generally, juice pulp reduction targets the range of a product with 6–12% pulp. Centrifuge efficiency in citrus plants for this process is simply calculation of the difference in percent pulp between the feed and product streams divided by the percent pulp in the feed stream. At normal feed rates, the process tries to achieve 20–50% pulp reduction. If the process requires greater than 50% reduction, slower feed rates and loss of capacity will result. It should be noted that since centrifuge applications result in juice yield loss, careful control and optimization is important.

The pulp reduction/clarification two-phase operation is diagrammed in the clarifier bowl schematic of Figure 4.6a. The juice–pulp feed stream enters at constant flow (pressure) from a surge tank into the top of the bowl and is conducted down to an area just outside the disks, where the two feed components are separated by the centripetal force of the rotating bowl. A timer or pressure control opens a valve allowing partial discharge of the solids at intervals to keep the centrifuge operating efficiently and the juice stream flowing smoothly up through the discs to the liquid outlet. The solids discharge frequency can be adjusted, with the objective of maintaining the discharged sludge as thick as possible. If the discharge is too liquid, yield loss will increase.

In applications where a feed stream contains liquids of different specific gravity as well as insoluble solids, a centrifuge is equipped with a separator bowl, diagrammed in Figure 4.6b. Recovery of cold-pressed oil from an oil/water/solids mixture is such a three-phase application. As in the clarifier, the feed stream is distributed down into the disks where the light phase (oil) and the heavy phase (water) are separated. The solids are conducted to the outside

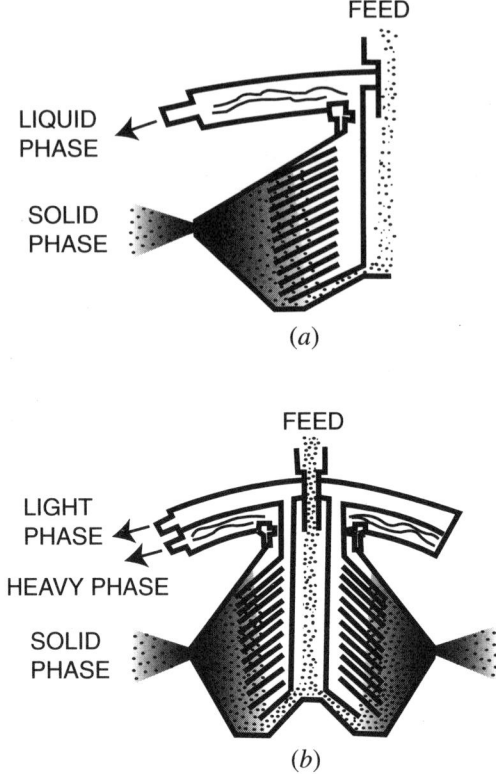

(a)

(b)

Figure 4.6 Diagrams of (a) continuous-centrifuge two-phase clarifier and (b) three-phase separator bowls for citrus processing applications.

of the bowl, where frequent discharge with some of the heavy phase material occurs as the heavy phase moves up to its discharge port. The lighter oil phase moves inward and up through narrow spaces in the disks to the discharge port. Citrus peel oil recovery is a two-centrifuge process: the first centrifuge, the desludger, concentrates the oil emulsion and the second, the polisher, breaks the concentrated emulsion, recovering the pure oil. More details of the peel oil recovery process are presented in Chapter 11.

The popularity of NFC juice coupled with the ability to store it in aseptic tank farms has resulted in a revival of an older citrus industry centrifuge process for lowering juice peel oil content. In this process, a centrifuge is used to remove some of the peel oil in the juice from the extractors. Claims are juice yield improvement, since the extractors can be set for maximum juice yield and quality comparable with low oil extracted juice. To prevent air from entering the juice stream and aroma loss during the process, the centrifugation is performed in hermetically sealed machines (Tetra Pak, 1998).

4.3.2 Summary

Centrifuge technology has taken a step forward with the development of vertical bowl assembly centrifuges that discharge pulp from juice in a continuous manner, without the bowl discharging features mentioned above. In one type of machine in citrus juice use, the pulp is discharged continuously and smoothly from the bottom section by a torque-controlled scroll, which allows concentration of the discharge at constant speed, increasing yield (Dorr-Oliver, 1997). A general discussion of these and other centrifuge operational variables and theory is available (De Loggio and Letki, 1994). Traditional centrifuges, which have to discharge portions of the bowl contents each time the machine activates a bowl cleaning cycle, lose rpm and yield. This can be optimized by frequent, minimal discharges rather than longer periods with complete discharge of the bowl contents.

REFERENCES

Bott, E. W. and Schöttler, P. (1989). Centrifuges, decanters and processing lines for the citrus industry, Tech. Doc. No. 4, Westfalia Separator AG, Oelde, Germany.

De Loggio, T. and Letki, A. (1994). New directions in centrifuging. *Chem. Eng.* **10**(1): 70–76.

Dorr-Oliver (1997). The ClariDry®, a new concept in centrifuges, Bulletin Clari-2, Dorr-Oliver Inc., Milford, CT.

FMC. (1993). Finisher automation, Product Bulletin 93-005, FMC Corp., Lakeland, FL.

FMC. (1998). Internet extractor information, http://www.fmcfoodtech.com.

Kimball, D. A. (1996). "Oranges and tangerines" and "Grapefruits, lemons and limes." In: *Processing fruits: Science and technology. Vol. 2. Major processed products*, L. L. Somogyi, D. M. Barrett, and Y. H. Hui, Eds., Technomic Publishing Co., Inc., Lancaster, PA. Chap. 9, pp. 265–304, Chap. 10, pp. 305–336.

Mendes, C. N. (1997). Configuration of a fruit juice extraction machine, U.S. Pat. 5,655,441.

Redd, J. B., Hendrix, D. L., and Hendrix, C. M., Jr. (1986). *Quality Control Manual for Citrus Processing Plants*, Vol I, Intercit, Safety Harbor, FL.

Tetra Pak. (1998). *The Orange Book.* Tetra Pak Processing Systems AB, Lund, Sweden. www.tetrapak.com. 206 pp.

Vitali, A. A. and Rao, M. A. (1984). Flow properties of low-pulp concentrated orange juice: Serum viscosity and effect of pulp content, *J. Food Sci.* **49**(3):876–881.

Waters, R. D., Cox, J. E., and Swofford, R. W. (1987). Method and apparatus for automatically controlling juice finishing machine, U.S. Pat. 4,665,816.

CHAPTER 5

SINGLE-STRENGTH JUICES AND CONCENTRATE

The juice recovery process determines whether the processing streams will be directed to single-strength NFC juice or to an evaporator for manufacture into concentrate. Common terminology and definitions of the various juice products named in this text are described on a web site (Terms, 1998). Except for a small amount of juice packaged and sold to consumers as fresh juice, most juice is thermally treated in heat exchangers or evaporators to kill spoilage microorganisms and inactivate native enzymes. Thermal treatment irreversibly alters the flavor of juice to the degree that it no longer has all the characteristics of fresh juice. For concentrate, the heat/vacuum process also removes aroma compounds essential to juice flavor, necessitating addition of flavor to the finished product in a blending step prior to packaging. This discussion will describe the pasteurization and concentration processes and their relationship to finished product quality.

5.1 FRESH JUICE

Because the aroma and flavor of freshly squeezed, nonpasteurized citrus juice is very desirable for the consumer, there is some local distribution of this product. The shelf life is very limited (<20 days at 1°C) and the potential for microbial spoilage is very high. The manufacturing operations from fruit washing to packaging must be exceptionally clean to minimize product spoilage and the product is kept as close to the freezing point as practical (Decio and Gherardi, 1992). Pectinesterase activity results in loss of cloudiness through serum/pulp separation. This is considered a defect in pasteurized juices; however, shrewd marketers describe this to consumers as proof that the product is pure and has never been heated to cause loss of important nutrients. Of course, the truth is that many important nutrients in citrus juice (water, sugar, acid, and some flavonoids, vitamins, minerals, and other compounds) are quite heat stable under the conditions of pasteurization. Some fresh product has traditionally been frozen for preservation; however, cloud separation, flavor changes due to reactions with oxygen, and color instability still occurs, although at a slower rate, than for product maintained cold but fluid (Joslyn and Marsh, 1933; Merin and Shomer, 1984).

Flavor compounds are sensitive to thermal degradation and pasteurization does result in loss of the so-called fresh juice flavor. Although initial fresh juice flavor is quite good, enzymes and microbial growth cause rapid deterioration, even under refrigeration. Thus, several days after packaging, flavors from di-acetyl, fusel oils, and other microbially generated off-flavors result in a product not equivalent to good-quality pasteurized juice, which may have a refrigerated shelf life of 6–8 weeks. In fact, it has been demonstrated that only a few hours are necessary for the sensory detection of off-aromas from microbial sources in nonpasteurized orange juice (Teller, 1993; Teller et al., 1992). A discussion of flavor properties of commercial fresh juice may be a moot issue due to outbreaks of food-borne illness from public consumption of nonpasteurized, packaged fruit juices. These include serious incidences of salmonellosis from consumption of contaminated fresh orange juice and an outbreak of *Escherichia coli* O157:H7 illness from nonpasteurized apple juice (Parish, 1997, 1998).

The Food and Drug Administration (FDA) has published a rule related to labels required on fresh, unpasteurized juice packages (FDA, 1998). The National Food Processors' Association (NFPA) believes the FDA should focus on providing the greatest protection possible for consumers, not on protecting the interests of a few unpasteurized juice producers. The FDA regulations on labeling of juice products do not require pasteurization or heat treatment for all juices, as NFPA has urged. The rule does require warning labels on unpasteurized products, to alert consumers that unpasteurized juices can contain bacteria that pose a special risk to certain individuals, particularly children and older adults. The FDA also has proposed juice regulations to mandate the use of Hazard Analysis and Critical Control Point (HACCP) by most juice producing companies. Procedures for fresh juice processors implementing HACCP programs have been published (Schmidt et al., 1997). However, if a product is pasteurized, HACCP is not necessary and adds burdensome and unnecessary requirements on the food industry (NFPA, 1998).

5.2   PASTEURIZED JUICE

Recent citrus juice consumption statistics indicate a trend to consumer preference for single-strength chilled and premium NFC products over concentrate, which must be diluted by the consumer before drinking (Nielsen, 1997). This trend is primarily due to the inconvenience of the thawing, mixing, and dilution procedure, although the flavor of the premium NFC products is excellent. Premium NFC juice represented 40% of the chilled single-strength juices for the Florida season ending in 1996, but only 30% for 1991. In 1991, FCOJ was 39% of total juice sales, decreasing to 27% by 1996. Necessity for food safety and quality requires pasteurization of these juices before packaging and distribution.

5.2.1 Enzyme Inactivation

Considering thermal kinetics of the pasteurized products, NFC juices should have the best flavor characteristics because they receive only one heat treatment. The reconstituted juices receive two thermal treatments, the preheating step of the concentration process and pasteurization after dilution and blending before packaging. For NFC citrus juices and the preheating step of evaporation, pasteurization is designed to adequately inactivate the thermally stable isozyme of pectinesterase (PE). The temperature necessary for this is approximately 90°C, which is 10–15°C higher than required to kill the microbes. This isozyme has about 10% of the total pectinesterase activity but must be inactivated for successful shelf life cloud stability. Pectinesterase also has a less thermally stable isozyme, which makes up about 80–90% of the total activity. Cameron and Grohmann (1996) have characterized the isozymes of Valencia orange juice pectinesterases and purified the thermally stable form of the enzyme. Pectinesterase is associated with the insoluble particulates in the juice and if not inactivated, causes the juice turbidity to break, allowing the cloud to separate from the serum. This is a serious quality defect, since consumers expect commercial citrus juices to be cloudy.

In designing pasteurization processes for microbial inactivation, D and z values are rate constants equating the inactivation curves to the temperature and time of the thermal treatment (Toledo, 1991). In microbiology, D refers to decimal value where $\frac{1}{10}$ of the initial organisms remain after a period of thermal treatment. Application of the concept to chemical and enzymatic reactions is possible, since the D value (measured in time units) is the negative reciprocal of the slope of the graph of log concentration (or activity) versus time (Fig. 5.1). The z value is the temperature change necessary for a factor of 10 (1 log cycle) reduction in enzymatic or microbial inactivation rate. The z value is related to the D value as the slope of the graph of log D versus temperature (Fig. 5.1). In chemical reactions, the z value is similar to the first-order reaction constant k, which varies with temperature and may be calculated from the activation energy by the Arrhenius equation (Toledo, 1991). For inactivation of the multiple forms of citrus juice pectinesterase, D and z values have been published (Eagerman and Rouse, 1976; Versteeg et al., 1980). Inactivation (99% = $2D$) of heat stable orange juice pectinesterase takes 0.8–1.0 min at 90°C, with 10-fold inactivation rate, z = 6.5°C. The unknown concentrations of both heat labile and heat stable pectinesterases in the various juices makes

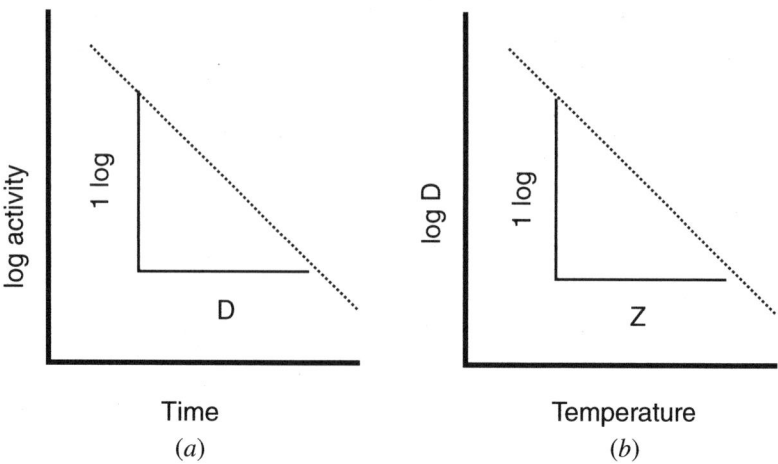

Figure 5.1   Graphs to illustrate calculation of (*a*) *D* value from log activity vs. time and (*b*) *z* value from log *D* value vs. temperature.

it difficult to use the above log-linear model for determining *D*-values for a juice. However, a model which incorporates both of these components offers a method for calculating a total juice *D*-value for pectinesterase (Chen and Wu, 1998).

Grapefruit juice has slightly lower thermal inactivation requirements, with values near 85°C for inactivation of the heat stable isozyme, $z = 6°C$ (Seymour et al., 1991). These values are not unique and are subject to variability depending on time of the season, amount of pulp in the juice, and type of fruit. There is also evidence that increasing acidity decreases pectinesterase activity. Wicker (1992) reported that pectinesterase extraction from grapefruit juice pulp at pH 3.0 and 8.0 resulted in constant thermostable isozyme activity. However, the heat labile isozyme, which makes up 90% of the total activity in juice, is sensitive to inactivation at low pH. Results have been published showing that holding at pH 2.0 for 5 min. is sufficient to inactivate the heat labile pectinesterase (Sun and Wicker, 1996).

While the above studies of the pH effect on pectinesterase activity are interesting, a relationship was established long ago for temperature inactivation at different pHs of this enzyme in orange, grapefruit, tangerine, and lime juices (Rouse and Atkins, 1953). These authors reported that by adjusting juice to below pH 3.5, even temperatures as low as 50°C resulted in 20–40% pectinesterase inactivation. Total inactivation at high temperatures (e.g., 90°C) could also be achieved in shorter times at lower pH. Rouse and Atkins didn't know at the time that the heat labile form of the enzyme was pH sensitive. Juice treatment with CO_2 at above supercritical conditions, where carbonic acid lowers juice pH and inactivates heat labile pectinesterase, followed by pressure release to attain previous juice atmospheric pH conditions has been patented

(Balaban et al., 1995). This process has the advantage of enzyme inactivation without heat, thus preserving the flavor, similar to the use of high pressure discussed in Chapter 6.

Pectinesterase activity is measured by a titrimetric procedure, usually conducted near pH 7, involving addition of a small amount of juice sample to a pectin solution. The enzyme de-esterifies the methoxyl groups of the pectin, producing methanol and pectic acid. The acid is titrated with base in a timed reaction, reporting the activity, expressed as PEU = microequivalents ester hydrolyzed/minute. In general, activity for cloud stable commercial refrigerated juices and frozen concentrates is less than 0.1–0.3 PEU. Pasteurized pulp and some very pulpy juices may have a higher activity of 0.5–1.0 PEU. Fresh unpasteurized orange juice, depending on the pulp content, may have an activity range of 1–10 PEU; while grapefruit juice is less, in the range of 0.5–5 PEU. Some early methods reported PEU/°Brix (Eagerman and Rouse, 1976; Redd et al., 1986), which causes some confusion in quality control procedures for citrus products, comparing old data with current results.

5.2.2 Aseptic NFC Juice

Chilled orange juice from concentrate is a convenient form of packaged juice readily available to consumers in the refrigerated sections in markets. However, as stated above, this product requires two thermal treatments, which are deleterious to the flavor. Also of importance to consumers is the requirement that the package labels contain the words, "From Concentrate." Currently, technology is available on a large scale to extract, process, and store NFC juice in bulk aseptic refrigerated tanks, minimizing microbial spoilage potential and product quality degradation (FMC, 1997). This technology enables providing blended juices to consumers on a year-round basis, when fruit is not being processed. The older process of freezing blocks of raw or pasteurized juice, storing the frozen blocks in freezer warehouses, followed by chopping and pasteurization for packaging has mostly been replaced with this convenient new technology. However, low freezer temperatures ($-20°C$) of the former process, preserved the chemical quality of the juice for periods of a year or more.

The technology for NFC bulk tank aseptic storage of juice is not too complex; however, because of the potential for large-scale economic loss if the system fails, it requires special care for implementation. Advantages of this process include system automation and inventory management, product blending and standardization capabilities, either bulk or packaged product filling under aseptic conditions, and reduced operation and maintenance costs. Depending on the processing capacity of a citrus plant, a number of tanks of capacity 950–3800 m^3 (250,000–1,000,000 gal) each are installed in refrigerated rooms or insulated with refrigeration. Tanks with 11,400 m^3 (3,000,000 gal) capacity are being studied. A typical commercial installation of 3800 m^3 (1,000,000 gal) tanks is illustrated in Figure 5.2. Such installations may have the capacity for storage of from 7560 to 95,000 m^3 (2 to 25 million gal) of

Figure 5.2 Commercial installation of an aseptic NFC citrus juice tank farm. (By permission of FMC/FranRica. Stockton, CA; Southern Gardens Citrus, Clewiston, FL).

NFC juice. The Florida industry capacity, by the end of the 1998 season, is estimated to be near 380,000 m³ (100 million gal) (Brocker, 1997).

The process description involves complete sterilization of the system from the juice pasteurizer, pumps, piping into and out of the tanks, and the tanks themselves. This technology has been described in detail (Raasch, 1996) and includes the following: (a) deaeration/deoiler coupled to (b) tubular heat exchanger/chiller, (c) piping, valves, pumps, and (d) bulk storage tanks. The entire process (a–d) is functional with a computer automated control system. Citrus juices have complex flavor profiles that are easily degraded by exposure to heat and oxygen. Process (a) is designed to remove oxygen and excess peel oils from the juice before the sterilization step. In general, juice is heated to 50–66°C (120–150°F), then flash-evaporated under vacuum to reduce the oxygen content and some volatile compounds. If oxygen reduction is the only objective, the highest vacuum achievable at the lower range of process temperatures at the shortest residence time is used. More efficiency is achieved at higher temperatures. The flash vapors are condensed first by water (the condensate added back to the juice stream) and next by a glycol chiller for oil and essence recovery. Oil recovery is by decantation in a chilled tank and the aqueous essence is returned to the juice stream. An alternative to the vacuum deoiling process (a) involves use of a centrifuge to lower the juice peel oil content. The O_2 level is generally reduced to <0.5 ppm and the volatile peel oil content to approximately 0.02% by these processes.

The sterilization process in (b–c) above must first depend on the ability to sterilize the vital parts of the heat exchanger, including the preheater, hold tube, chiller, and all downstream components through the bulk tank system and its inlet/outlet valve manifold system. The heat exchanger system is sterilized with hot water at 121°C (250°F) for 30 min prior to product introduction. Sterilization of the bulk tanks (d) is accomplished by rinsing with potable water, caustic, and phosphoric acid. Then the tanks are filled with an iodofor solution for sterilization. Prior to product introduction, the piping connecting the tanks to the heat exchanger system is sterilized with hot water. Conditions for sterilization of the juice are, in part, dictated by conditions for inactivation of the heat stable pectinesterase described above. However, there is a need for some margin of error due to differing juice variables, such as variety, maturity, and pulp content. An aseptic process for orange juice would provide for product at 90–99°C (194–210°F) for 1–6 sec, based on pectinesterase inactivation kinetics. Longer hold times to 15–30 sec may be normal in actual commercial practice. The sterile product is chilled to a temperature near 2°C (35°F) and filled into the aseptic storage tank system.

Smaller aseptic tanks (to 950 m^3 capacity) are polished stainless steel. The larger tanks (to >3800 m^3), for strength, are constructed of carbon steel and have internal epoxy liners (coatings) to provide for chemical inertness and sterilizability. Reaction of the epoxy with d-limonene and subsequent potential for flavor scalping are negligible. The cost of juice processing and storage depends on the size of the installation, decreasing with the larger sizes to near 0.5 cents/L/yr (2.0 cents/gal). With proper nitrogen blanketing and mixing, juice quality may be maintained for a year or more. With proper maintenance, it has been reported that tanks have remained sterile for over 4 years without having to be cleaned and resterilized (Raasch, 1996).

5.2.3   Single-Strength Juice from Concentrate

Because of the economics of storing large bulk quantities of concentrated citrus juice and the consumer preference for a ready-to-serve product, a significant amount of orange juice is packaged from reconstituted concentrate as chilled juice (COJ). Although NFC juice volume is increasing, the volume of COJ is greater for now. The process for manufacture of COJ involves the following steps. The finished juice is thermally treated in the evaporator to inactivate enzymes and microbes, and concentrated to 65 °Brix. The evaporator pump-out is chilled and pumped to bulk tank farm (nonaseptic) storage. Next, the concentrate is pumped from the tank farm to a blend room, where water, essence, flavor, pulp, and so forth is added for reconstitution to the preferred single-strength juice product. This blending process may be computer controlled to allow for precision, product quantitation, and record keeping. The final steps are pasteurization at near 80°C for 15–30 sec to kill microbes, followed by packaging in cartons or glass. The high temperatures required for PE inactivation are unnecessary, since the enzyme should have been inactivated during the concentration process. The flavor of COJ may not be of the quality of either NFC or FCOJ because of the two thermal treatments and the loss of volatiles during the concentration process. However, significant advances in flavor chemistry have improved the sensory properties of the added aromas and essences and resulted in continual improvements in quality of this blended product.

The 65 °Brix concentrate may be shipped in bulk truck tankers to other citrus processors for retail packaging or to a port tank farm facility to be loaded onto ocean-going tankers for export. In the latter case, the ultimate packager may be a juice, beverage, or a dairy plant, which packages citrus juices. Port facilities for exporting and importing 65 °Brix bulk concentrate are available in Brazil, the United States, and Europe.

5.3 FROZEN CONCENTRATED JUICES

Much has been written about concentrated citrus juices in other sources. This section is written to help secure the future importance of these products to the industry and describe the process in view of the potential for by-products recovery. The primary product is FCOJ, with all of the other commercial citrus juices also processed into concentrates. The primary water removal technology is high-temperature short-time evaporation, although freeze concentration and membrane processes are also used.

5.3.1 Manufacture

Citrus juice evaporators are designed to heat juice rapidly for as little time as necessary to inactivate pectinesterase, then flash the juice through the stages under vacuum to produce the concentrate. Product flow is feed-forward through the stages (from lowest to highest °Brix), while the effects (where steam vapor from the product is condensed) may be arranged for maximum energy efficiency. Detailed descriptions of typical citrus evaporators and their operation have been published (Chen et al., 1993). They may have capacities from 14,000 to 91,000 kg (30,000 to 200,000 lb) water evaporated/hour, up to 7–8 stages and 4–7 effects. The process flow pattern involves preheating the juice to 90–100°C (194–212°F). This inactivates the pectinesterase, preventing gel formation in the stored concentrate. The preheated juice is flashed under vacuum in the top of the first stage tube nest. Juice (10–12 °Brix) then flows in a thin film down through the tubes to the bottom, is pumped to the top of the next stage, and so forth, to the final stage, where it is pumped out of the evaporator at 65 °Brix. As the juice water is evaporated in each effect, the vapor travels up the tubes to become the steam to heat the juice in the next effect. Volatile aromas and oils in the juice are stripped off in the vapor streams with the water and recovered by chilled water condensing and rectification in a small still (essence recovery unit) near the top of the evaporator. Only the highest concentration vapors containing the most valuable flavor notes are condensed for aroma recovery. These usually include only those vapors from the lower temperature stages.

Because of the long narrow tubes and high fluid velocities required for efficient evaporator operation, juice viscosity must be kept to a minimum. The serum viscosity of the single-strength juice entering the evaporator will be increased by a power function as the concentration increases in successive effects of the evaporator (Vitali and Rao, 1984a,b). Reduction and control of the initial juice pulp content by centrifugation helps maintain lower viscosity in the concentrated juice, both in the evaporator and at the low temperatures of concentrate storage. An alternate effective technique used in the processing industry for reducing concentrate viscosity has been the placement of homogenizers either within or just after the evaporator (Grant, 1989).

The 65 °Brix evaporator pump-out is blended with a small amount (<0.01% v/v) of cold-pressed peel oil to mask off-flavors that develop in storage, passed through a heat exchanger for chilling to the storage temperature of $-9°C$ (15°F) and pumped to the tank farm. Concentrate tank farms consist of 200–950 m³ (50,000–250,000 gal) stainless steel tanks kept in a cold room at the storage temperature. Large processing plants may have storage capacity >76,000 m³ (20 million gal) 65 °Brix concentrate. This concentrate is blanketed with nitrogen and carefully monitored for quality characteristics, so that juice of different characteristics may be accurately blended to produce finished products. Concentrate may be maintained in these tanks for over a year with little quality loss.

A problem may occur with crystallization of potassium citrate crystals in the 65 °Brix concentrate. These crystals have the appearance of sand and contribute to defects in the juice, since they can co-precipitate with hesperidin and become less soluble at time of reconstitution. Blending and very low temperature storage are offered as solutions to preventing formation of these crystals, as formation is a thermodynamic process (Kimball, 1985). At the time of blending to make single-strength juice or FCOJ, proprietary flavor materials composed of aromas, peel oils, pasteurized juices, and pulp may be added, depending on the product. Following the blending process, product is packaged for retail distribution, either as FCOJ or COJ.

Evaporator operation and juice quality also depend on keeping the evaporator clean. Hesperidin in the juice will form scale on the juice side of the tubes, decreasing efficiency of the evaporator, resulting in increased temperature to maintain capacity. This can result in a cooked flavor in the concentrate. Periodic cleaning involves water rinsing, cleaning with 2–5% caustic at normal steam settings, followed by rinsing to remove residual caustic (Redd et al., 1992). The caustic is recycled for economic and environmental reasons. The caustic stream from the evaporator will be bright yellow from the dissolved hesperidin and other flavonoids (Chapter 15).

5.3.2 Evaporator Efficiency

Some common questions related to concentrate manufacture by evaporation are dealt with in the following discussion. There is occasionally a need to calculate the capacity of a juice evaporator to determine if it is operating efficiently. The following example will illustrate how to calculate evaporator capacity from equations in Chen et al. (1993). Since evaporators are commonly rated in pounds of water removed/hour, the following calculations are in English units.

First determine the single-strength juice feed rate to the evaporator (e.g., 140 gal/min of 12 °Brix juice). The evaporator product is 65 °Brix concentrate. Next refer to Brix tables for densities. For 12 °Brix: density = 8.724 lb/gal; soluble solids = 1.047 lb solids/gal. For 65 °Brix: density = 10.977 lb/gal; soluble solids = 7.135 lb solids/gal. In an evaporator, solids in = solids out, since only water is removed. For 12 °Brix: Solids in = 140 gal/min $\times$ 1.047 lb solids/gal = 146.6 lb solids/min. For 65 °Brix: Solids out = X gal/min $\times$ 7.135 lb solids/gal = 146.6 lb solids/min; X = 146.6 lb solids/min $\div$ 7.135 lb solids/gal = 20.5 gal/min of 65 °Brix product. Water removed in the evaporator = wt. feed in $-$ wt. product out = 140 gal/min $\times$ 8.724 lb/gal $-$ 20.5 gal/min $\times$ 10.977 lb/gal. Thus, by calculation:

Evaporator capacity = 996 lb/min $\times$ 60 min/hr = 59,780 lb water/hr.

Once the evaporator capacity has been determined, for production reasons, the juice feed rate of a particular °Brix juice may be required. A quick method is as follows:

Juice feed rate = evaporation capacity $\div$ [1 $-$ °Brix in/°Brix out]

A 100,000 lb/hr capacity evaporator would have a juice feed as follows:

Feed = 100,000 lb/hr $\div$ [1 $-$ 12/65] = 122,642 lb/hr (14,058 gal/hr)

At a juice yield of 50-lb juice/90-lb box of oranges, this 100,000 lb/hr evaporator can concentrate the juice from 2453 boxes of oranges/hr.

5.3.3   Concentrate Dilution

Another typical concentration problem involves adjusting the °Brix of a concentrate by dilution with either water or single-strength juice. A common material balance method utilizing algebra, taught in chemistry classes, may be used if Brix tables are available. Some useful basic computer programs have also been published for this purpose (Kimball, 1991). The following example is a typical problem illustrating how °Brix tables may be used to dilute concentrate to a single-strength product:

Example   Dilute 1 gal of 58 °Brix concentrate to 44.8 °Brix with 12 °Brix juice.

Conc. $\times$ vol. (in) = conc. $\times$ vol. (out)

Blend 58 °Brix + 12 °Brix to obtain 44.8 °Brix product

$$6.165 \text{ lb/gal} \times 1 \text{ gal} + 1.047 \text{ lb/gal} \times X \text{ gal} = 4.488 \text{ lb/gal} \times Y$$

Solids balance: $6.165 \times 1 + 1.047 \times X = 4.488 \times Y$

Total materials: $1 + X = Y$

Solving: $X = 0.49$ gal 12 °Brix $Y = 1.49$ gal 44.8 °Brix

Most evaporators are computer controlled based on such sensors as in-line refractometers or true mass flow meters measuring liquid concentrations and fluid densities in certain effects as well as the juice flow rate entering or product leaving the evaporator. Some schemes may also measure temperatures for control purposes. Automation compared with manual control saves energy, allows longer operation before cleaning, and improves concentrate quality by narrowing the deviations from equilibrium brought about by such external factors as feed and steam flow variations. These systems may also be tied in to the whole plant control system, as in the juice control room. Most juice processors work cooperatively with equipment and evaporator manufacturers to design an automation scheme suitable to the plant's requirements and capabilities.

REFERENCES

Balaban, M. O., Marshall, M. R., and Wicker, L. (1995). Inactivation of enzymes in foods with pressurized CO_2. U.S. Patent 5,393,547.

Brocker, P. (1997). FMC/FranRica, private communication, Paul—brocker@fmc.com.

Cameron, R. G. and Grohmann, K. (1996). Purification and characterization of a thermally tolerant pectin methylesterase from a commercial Valencia fresh frozen orange juice, *J. Agric Food Chem.* **44**(2):458–462.

Chen, C. S. and Wu, M. C. (1998). Kinetic models for thermal inactivation of multiple pectinesterases in citrus juices, *J. Food Sci.* **63**(5):747–750.

Chen, C. S., Shaw, P. E., and Parish, M. E. (1993). "Orange and tangerine juices." In *Fruit Juice Processing Technology*, S. Nagy, C. S. Chen, and P. E. Shaw, Eds., AgScience, Auburndale, FL, Chap. 4.

Decio, P. and Gherardi, S. (1992). Frischaft aus Orangen, *Confructa Studien* **36**(5/6): 162–167.

Eagerman, B. A. and Rouse, A. H. (1976). Heat inactivation temperature–time relationships for pectinesterase inactivation in citrus juices, *J. Food Sci.* **41**:1396–1397.

FDA (1998). Final rule on FDA labeling of fresh juice, *Fed. Reg.* July 8, 37029–37056.

FMC (1997). http://www.FMCFMG.com/FranRica.

Grant, P. M. (1989). Citrus juice concentrate processor, U.S. Pat. 4,886,574.

Joslyn, M. A. and Marsh, G. L. (1933). Frozen orange juice. Its manufacture, preservation and distribution, *Canning Age* **14**(4):229–230, 235–236, 238.

Kimball, D. A. (1985). Crystallization of potassium citrate salts in citrus concentrates, *Food Technol.* **39**(9):76–81, 97.

Kimball, D. A. (1991). *Citrus Processing Quality Control and Technology*, Van Nostrand Reinhold, New York.

Merin U. and Shomer, I. (1984). Structural stability of fresh and frozen-thawed Valencia (*C. sinensis*) orange juice, *J. Food Sci.* **49**(6):1489–1493, 1512.

NFPA (1998). http://www.nfpa-food.org.

Nielsen, A. C. (1997). *Retail OJ Sales*, Florida Dept. Citrus Reference Book, http://www.fred.ifas.ufl.edu/citrus/index/html.

Parish, M. E. (1997). Public health and non-pasteurized fruit juices, *CRC Crit. Rev. Microbiol.* **23**(2):109–119.

Parish, M. E. (1998). Coliforms, *Escherichia coli* and *Salmonella* serovars associated with a citrus processing facility in a Salmonellosis outbreak, *J. Food Protect.* **61**(3): 280–284.

Raasch, J. B. (1996). Aseptic processing and storage of citrus juices, *Trans. Citrus Eng. Conf.* **42**:71–80.

Redd, J. B., Hendrix, D. L., and Hendrix, C. M., Jr. (1986). *Quality Control Manual for Citrus Processing Plants*, Vol. I, Intercit, Safety Harbor, FL.

Redd, J. B., Hendrix, D. L., and Hendrix, C. M., Jr. (1992). *Clean-up Procedures for TASTE Evaporator*, Vol. II, AgScience, Auburndale, FL, pp. 141–143.

Rouse, A. H. and Atkins, C. D. (1953). Further results from a study on heat inactivation of pectinesterase in citrus juices, *Food Technol.* **7**(6):221–223.

Schmidt, R. H., Sims, C. A., Parish, M. E., Pao, S., and Ismail, M. A. (1997). "A model HAACP plan for small-scale, fresh-squeezed (not-pasteurized) citrus juice operations." Univ. FL Coop. Ext. Serv. Cir. No. 1179. Gainesville, FL, 20 pp.

Seymour, T. A., Preston, J. F., Wicker, L., Lindsay, J. A., Wei, C. I., and Marshall, M. R. (1991). Stability of pectinesterases of Marsh white grapefruit pulp, *J. Agric. Food Chem.* **39**:1075–1079.

Sun, D. and Wicker, L. (1996). pH affects Marsh grapefruit pectinesterase stability and conformation, *J. Agric. Food Chem.* **44**:3741–3745.

Teller, H. K. (1993). Microbial effects on flavors in orange juice, PhD Dissertation, University of Florida. Gainesville, FL.

Teller, H. K., Parish, M. E., and Braddock, R. J. (1992). Microbially produced off-flavors in orange juice, *Proc. Fl. State Hort. Soc.* **105**:144–146.

Terms (1998). Glossary of juicy terms, www.fred.ifas.ufl.edu/citrus.

Toledo, R. T. (1991). "Kinetics of chemical reactions in foods." *Fundamentals of Food Process Engineering*, 2nd ed., Van Nostrand Reinhold. New York, Chap. 8, pp. 310–314.

Versteeg, C., Rombouts, F. M., Spaansen, C. H., and Pilnik, W. (1980). Thermostability and orange juice cloud destabilizing properties of multiple pectinesterases from orange, *J. Food Sci.* **45**:969–971, 998.

Vitali, A. A. and Rao, M. A. (1984a). Flow properties of low-pulp concentrated orange juice: Serum viscosity and effect of pulp content, *J. Food Sci.* **49**(3):876–881.

Vitali, A. A. and Rao, M. A. (1984b). Effect of temperature and concentration, **49**(3): 882–888.

Wicker, L. (1992). Selective extraction of thermostable pectinesterase, *J. Food Sci.* **57**(2):534–535.

CHAPTER 6

JUICE CHEMICAL REACTIONS, PRODUCT STABILITY, AND PACKAGING

Orange juice is dominant among the various citrus juices; therefore, much of the available information about quality and stability deals with this product. The basic flavor differences among the various citrus juices relate to sweetness, tartness, terpene, and other volatile aromas characteristic of each type of fruit. Initial product quality is dependent on the conditions of processing and variables associated with the biochemistry of the fruit. Once packaged, storage environments affecting the chemical composition of juice components determine the maintenance of high quality. In this environment, storage temperature is the single most important variable related to product stability, followed by presence of oxygen, nonoxidative reactions, and interactions of flavors with packaging materials.

6.1 CHEMICAL REACTIONS

The flavor and aroma of freshly squeezed nonpasteurized citrus juice is the target for the optimum initial quality of pasteurized, packaged products. Irreversible damage to the flavor results from chemical reactions initiated or occurring during the heating process. Other compounds, such as vitamin C, react kinetically during heating and after packaging to result in quality loss. In many cases, the degradative reactions involve oxygen dissolved in the product prior to processing or diffusing through the package into the product during its shelf life. The quality requirement for maintaining a stable cloudiness in citrus juices necessitates that juice be subjected to a significant amount of heat to inactivate the enzyme responsible for turbidity (cloud) loss. This thermal treatment, in part, is responsible for initial chemical and flavor degradation of the juice.

6.1.1 Pectinesterase

Enzyme reactions are considered within the realm of biochemistry and are most important in fresh, nonpasteurized juice. As discussed in the last chapter, pectinesterase is the enzyme of primary concern for citrus juice product stability. The chemical reaction products of pectinesterase with its juice substrate, pectin, are methanol and pectic acid and result in the irreversible destabilization of the cloud. If allowed to occur in the raw juice, cloud loss carries through into subsequently pasteurized products. Enzyme reaction rates all depend on the concentrations of enzyme and substrate and occur at optimal pH and temperature conditions. Enzyme reaction rates are sometimes thought of in terms of "turnover number" or the velocity of conversion of units of substrate to products, and values may range from as few as 100 to over a million/sec. In general, esterases (e.g., pectinesterase) have turnover numbers of 10,000/sec. The pectinesterase reaction in raw juice is very rapid, and, if the final product cloud is to be stable, it is important to pasteurize the juice as quickly as possible after extraction and finishing.

Pectinesterase concentrations, determined by activities in various tissues of the commercially important fruit, have been studied extensively because of the importance of cloud loss. As can been seen from examination of Table 6.1, the majority of the enzyme is present in the internal fruit tissues, the juice pulp, and rag membrane. For example, albedo from Valencia oranges only has about 9% of the activity of the juice pulp. These results should alleviate any concerns about significant quantities of pectinesterase from the peel (flavedo and albedo) getting into the juice during extraction, as suggested by some studies (Cameron et al., 1997, 1998). More significantly, treatments such as harsh extraction or finishing that disrupt the internal tissue (pulp and rag) in the extracted juice may affect the activity. This is especially so because pectinesterase is an enzyme that is bound ionically to the membranous tissue and not freely soluble in the juice (Jansen et al., 1960).

TABLE 6.1 Activity of Pectinesterase (PEU) in the Component Parts of Oranges and Grapefruits[a]

	PEU/g component $\times$ 10^3				
	Flavedo	Albedo	Rag	Pulp	Juice[b]
Hamlin	15.6	10.6	22.8	55.3	0.1
Pineapple	14.6	8.5	23.8	129.9	0.4
Valencia	9.5	6.6	14.1	76.0	0.2
Navel[c]	7.0	4.8	25.5	45.7	—
Grapefruit	17.2	12.8	6.9	30.0	0.2

[a]Adapted from Rouse (1953).
[b]Activity in juice centrifuged free of insoluble solids.
[c]Rombouts et al. (1982).

Citrus juice cloud stability is a complex process, primarily associated with pectinesterase activity. Juice pectin acts as a colloidal stabilizer that gives juice body and acts to suspend the cloud, which is composed of lipids, terpenes, protein, and polysaccharides (Scott et al., 1965). Cloud loss in juice is dependent on the presence of both pectic acid and divalent cations, such as Ca^{2+} (Krop and Pilnik, 1974). The chemical composition of orange juice cloud has been studied more recently by Klavons et al. (1991), who determined that insolubility of cloud protein was important in destabilizing the haze. Hesperidin has also been proposed as a constituent necessary for stable cloud (Ben-Shalom and Pinto, 1986).

Cloud stability for orange, grapefruit, and lemon juice was related to residual pectinesterase activity, acidity, and the degree of concentration (Rothschild and Karsenty, 1974). These authors reported that at 5°C, juices with 5% or less of the activity of fresh juice were cloud stable for many months. However, juice with high pectinesterase activity (1–2 units) typical of fresh juice was cloud stable only on the order of hours. Cloud destabilization has also been related to multiple pectinesterase forms, each with a different thermal stability and role in cloud loss (Versteeg et al., 1980), and to other factors, flavonoids, pH, and microbial contamination (Baker and Cameron, 1999).

Under certain conditions, addition of commercial pectolytic enzymes to raw juice may even confer cloud stability without the need for heating. An explanation of such a cloud stabilization mechanism concluded that enzymatically produced pectic acid hydrolysates, with a degree of polymerization from 8–15, inhibited pectinesterase in the juice (Termote et al., 1977). Juice with stabile cloud has been defined as having less than 36% light transmittance at 650 nm after a centrifugation step (Redd et al., 1986).

6.1.2   Pectin

The flow properties and mouth-feel of citrus juice products are strongly influenced by the amount of pulpy material and by the composition of the pectic substances. The importance of pectin to the juice rheological properties of viscosity and gelation have been reviewed by Ranganna et al. (1983a). There are three important pectin fractions defined by successive extractions, followed by precipitation with alcohol and additional extraction of the precipitate. These have been defined as water-soluble, ammonium-oxalate-soluble, and sodium-hydroxide-soluble pectins (Rouse et al., 1962). These pectin fractions are very important to juice and concentrate processing, as well as the properties of beverages and drinks manufactured from concentrates.

6.1.2.1   Orange Juice   The so-called water-soluble pectin contributes greatly to the juice viscosity and cloud stability. High amounts of this fraction can result in very high viscosity in concentrates and problems in operation of the evaporator and handling the product. It has been shown that increased juice pulp content increased apparent viscosity in 65 °Brix concentrate (Vitali and

Rao, 1984). In attempts to decrease concentrate viscosity, some industry evaporators have been fitted with in-line homogenizers, which have been proven to be at least partially effective (Grant, 1989). During heating or storage, hydrolysis may increase the water-soluble pectin from the pulp and insoluble particles in the juice. In a study reporting quality attributes of Valencia orange juice from light- and hard-squeeze extractor settings, the water-soluble pectin fraction doubled (from 570 to 1180 mg/L) under hard-squeeze settings (Attaway and Carter, 1975). This fraction also increases seasonally with fruit maturation and is historically higher in midseason fruit.

The oxalate-soluble pectin fraction is the insoluble pectate salt formed by reaction of Ca^{2+} ions in the juice with the pectic acid formed by pectinesterase de-esterification. This fraction may be indicative of pectinesterase activity in juice prior to pasteurization and results in two defects, potential gelation of concentrate and cloud loss. Sodium-hydroxide-soluble pectin is generally the total pectin in the juice and includes the protopectin from cellular fragments such as pulp and membrane. For quality monitoring purposes, commercial FCOJ (diluted to 11.8 °Brix) may contain water-soluble pectin (400–2000 mg/L), oxalate-soluble pectin (400–1000 mg/L), and sodium-hydroxide-soluble pectin (800–3000 mg/L) (author's unpublished data).

6.1.2.2 Grapefruit Juice   Grapefruit juice contains more pectin than orange juice. Rouse et al. (1965) reported that seasonal averages of total pectin (sodium hydroxide soluble) in grapefruit juice ranged from 2000 to 6000 mg/L, with the water-soluble pectin about 75% of the total. The oxalate-soluble fraction would be estimated to be the remaining 25% of the total. The greater amounts of water-soluble pectin in grapefruit juice influence the viscosity and may result in achieving lower concentrations when the juice is evaporated. There is considerable discrepancy in the above pectin substance values with some published data. For instance, the following identical maximum values are reported (RSK-Values, 1987) for water-soluble, oxalate-soluble, and alkali-soluble pectins, respectively, in both orange and grapefruit juice: 500, 200, and 300 mg/L. The author's experience supports the results mentioned from Attaway and Carter (1975) and Rouse et al. (1965). Perhaps the RSK data contain errors of methodology or interpretation because it is unlikely that the concentration of water-soluble pectin is greater than alkali-soluble pectin. For orange and grapefruit juices, variables such as time of the season, extraction, finishing, centrifugation, and homogenization may significantly alter these values.

6.1.3 Bitter Compounds

There are two classes of bitter chemicals in citrus juices: limonoids and flavonoids. The detrimental action of these compounds to acceptable juice flavor generated some early publications, to which the reader may refer (Guadagni et al., 1974; Horowitz and Gentili, 1977). Recently, the potential exists for juice bitterness reduction through genetic engineering to develop hybrids with nonbitter character or very low levels of these bitter compounds (Hsu et al., 1998). The following discussion relates the influence of these compounds to juice processing. The by-products potential of bitter principles from citrus is considered in Chapter 16.

6.1.3.1 Limonin The primary bitter limonoid detrimental to citrus juice quality is the compound limonin, a tetranortriterpenoid found mostly in the seeds and tissue of the fruit. Bitterness from this compound in citrus juices is mostly a problem for Navel oranges and grapefruits. High limonin levels are also found in the fruit from Sour Orange rootstock. Juices from the more commonly processed sweet oranges, such as Hamlin, Pineapple, and Valencia varieties have lower levels of limonin, not exceeding bitterness thresholds at fruit maturity. Limonin has poor water solubility, about 5 mg/L at 25°C and near 50 mg/L at >85°C, with a tendency for the molecule to undergo hydrolysis to the more soluble hydroxyacid lactone at higher temperatures (Chandler and Robertson, 1983). Juice and tissue also naturally contain some more water-soluble sugar–limonin derivatives, limonoid glucosides (Fig. 6.1), which contribute slightly to juice bitterness and may have anticancer properties (Hasegawa et al., 1989).

The sensory threshold of limonin in water has been established at between 3 and 6 mg/L; while the detection level in juice may be higher. Another limonoid, nomilin, is also as bitter as limonin and may contribute to the total bitterness sensation. Nomilin concentration in early-season grapefruit juice may be as high as 40% of the total limonoids; thus, its contribution to bitterness is significant (Rouseff and Matthews, 1984). Detection of these compounds in juice is necessary for some quality measures, and chemists have developed some analytical methods useful for this purpose. One rapid test is based on an antigen–antibody reaction of limonin (Jourdan et al., 1984; Ram et al., 1988), while HPLC methods are used in some citrus laboratories (Shaw and Wilson, 1988). Limonin decreases significantly as the fruit matures, which has the prac-

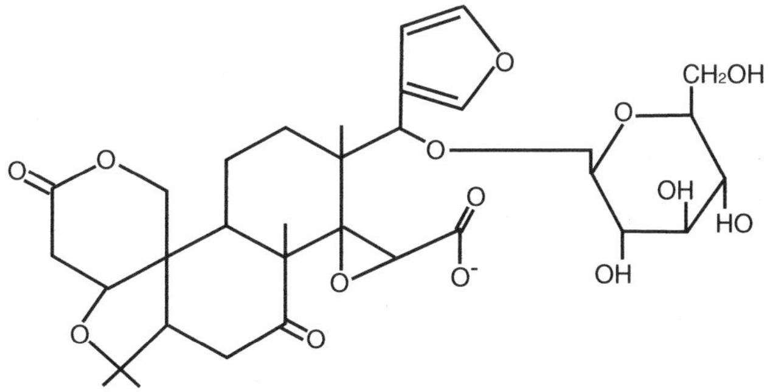

Figure 6.1 Structure of a limonin glucoside (limonin 17-β-D-glucoside) found in citrus fruit tissue.

tical consequence that undesirable bitterness may disappear as the processing season progresses. This is true primarily for early-season grapefruits, as navel oranges have such high amounts of limonin that the decrease does not effectively improve the flavor. Correlation of the amount of bitterness in various orange and grapefruit juices with processing parameters, such as different plants, juice pulp content, and time of season, gave inconsistent results for pulp content but decreased with maturity (Albach et al., 1981).

The chemistry of limonin involves conversion of the nonbitter limonin hydroxyacid lactone in fruit tissue to bitter limonin (Fig. 6.2). Chemical structures and a discussion of the biochemistry of these molecules related to their taste properties have been presented (Hasegawa and Maier, 1983). It is also possible to add enzymes to juice, thereby reducing bitterness. If one eats a freshly peeled fruit or drinks freshly squeezed navel orange juice, the bitterness is not detected. Upon pasteurization, heat, in combination with the acid in the juice, closes the D-ring to form the lactone, limonin, which is intensely bitter. In the case of fresh, nonpasteurized juice, low-temperature storage after packaging

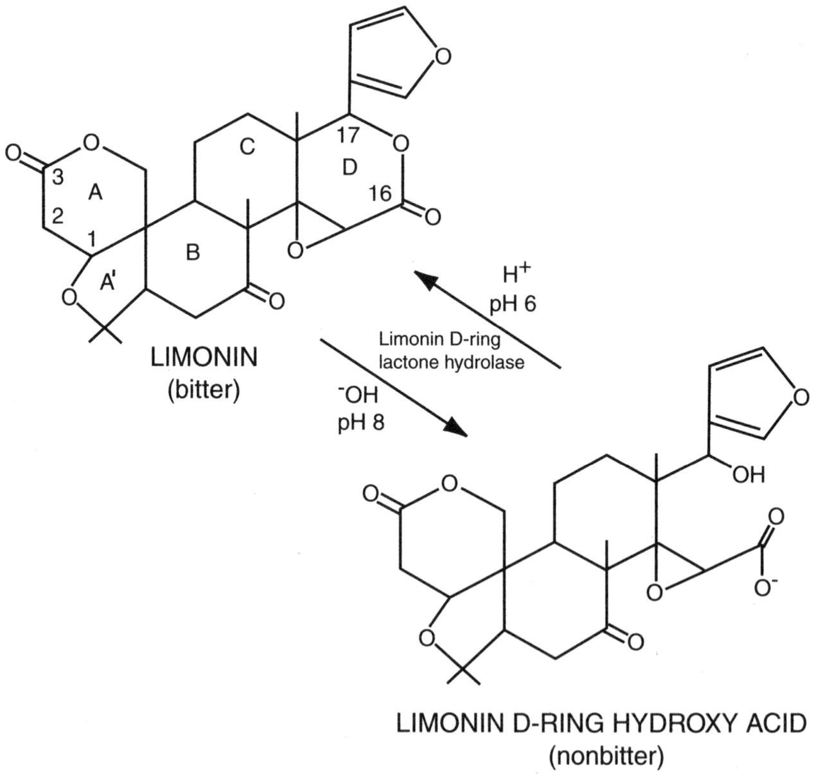

Figure 6.2   Reaction of bitter and nonbitter forms of limonin due to heat, acid, and enzyme hydrolysis in citrus juice.

may only delay the onset of bitterness by a matter of hours. Although the ring closure rate is slower at low temperature, it is still catalyzed by the juice acidity.

6.1.3.2 Naringin Bitterness is a major problem with consumer preference for grapefruit juice, which may suffer not only from limonin-caused bitterness but bitterness resulting from the flavonoid, naringin. The problem is compounded by the removal of bitterness because this flavor characteristic is typical in grapefruit juice. Fellers et al. (1986) have shown that consumers may even confuse bitterness and tartness in grapefruit juice. A recent study suggested that certain people were more sensitive to naringin bitterness and that increased acuity to bitterness corresponded to increased dislike for bitter naringin solutions (Drewnowski et al., 1997). Like limonin, naringin is mostly in the tissue of the fruit and pulpy juice material and also has a nonbitter isomer. Naringin is esterified with the disaccharide neohesperidose (glucose + rhamnose in the 2-position), which confers bitterness. Hydrolysis of neohesperidose from the molecule yields naringinin, which is nonbitter. The nonbitter naringin isomer is esterified with the sugar, rutinose (glucose + rhamnose in the 6-position). Hydrolysis of the rhamnose only, leaving the glucose yields the less-bitter compound, prunin. The structure and chemistry of these molecules has been reviewed and published (Horowitz and Gentili, 1977).

Aqueous solubility of naringin is considerably different than limonin. In water, the solubility increases asymptotically with temperature, from 500 mg/L at 20°C to greater than 100 g/L at 75°C (Pulley, 1936). Other than its solubility, the molecule is quite heat stable. Because of its solubility, it may be extracted from tissue or pulp particles in the juice during pasteurization or storage, resulting in increased juice bitterness. Naringin content of grapefruit peel and juice decreases as the fruit matures, leading to the knowledge that bitterness decreases dramatically if the fruit is allowed to ripen properly. The sensory threshold for detection of naringin bitterness in water was reported to be about 20 mg/L (Poore, 1934). A survey of the quantities of common flavonoids in 52 citrus cultivars reported that naringin was only present in grapefruit, sour orange, and some grapefruit hybrids, ranging from 100 to 800 ppm in grapefruit juice (Rouseff et al., 1987).

6.1.4 Juice Bitterness Reduction

6.1.4.1 Enzyme Methods Reduction of limonin bitterness in navel orange juice through the use of enzymes or biochemistry has been primarily a subject of research. An early report cited the need for enzyme research, debittering navel orange juice and general metabolic control of fruit enzymes with additives (e.g., diisofluorophosphate) (Balls, 1949). Biochemical approaches to this problem have been described, with attempts to reduce bitterness in the fruit, prior to juice extraction (Maier et al., 1971). Except for hydrolases (e.g., pectinases), there has been very little potential for commercialization of enzyme reactions in food processing. In general, enzymes are expensive and have limited availability when considering an application as specific as juice debittering by limonin reduction. One approach involved the addition of an enzyme isolated from *Pseudomonas sp.*, limonoate oxidoreductase, to bitter juices and products (Hasegawa et al., 1975). This enzyme reduced bitterness through formation of nonbitter dehydrolimonoic acid lactone; however, the process involved adjusting the juice pH. A review of the enzymatic debittering of juices also has reported that albedo tissue from orange peel contains enzymes with some ability to degrade limonin (Nicol and Chandler, 1978). Another review also has considered this subject (Ranganna et al., 1983b). It has been shown that grapefruit albedo containing limonin and naringin responded to gibberellic acid treatment by biologically reduced concentrations of these bitter chemicals (Shaw et al., 1991). Other research reported that immobilized cells and an enzyme complex from *Corynebacterium fascians* can metabolize limonin and nomilin in juice, resulting in reduced bitterness (Herman et al., 1985; Hasegawa et al., 1985).

Enzyme use to reduce bitterness of grapefruit juice by reducing the naringin concentration has, like limonin, been a subject of research. Some early studies found that enzyme contaminants in commercial pectinases had the ability to reduce grapefruit juice bitterness by hydrolysis of naringin (Ting, 1958). Subsequently, enzyme manufacturers screened a number of microorganisms used in pectinase manufacture for their ability to produce naringinase (Thomas et al., 1958). Some enzyme fractions with the ability to hydrolyze naringin and prunin were also purified from a commercial naringinase (Dunlap et al., 1962). Enzymatic hydrolysis reduced grapefruit juice bitterness from both naringin and narirutin; however, cloud loss occurred during enzyme reaction time (Versteeg et al., 1977). Immobilization of naringinase in dextran (Ono et al., 1977) and alginic acid gels (Puri et al., 1996) may allow more efficient contact with grapefruit juice than direct enzyme addition. Naringinase has also been immobilized in the lumens of ultrafiltration hollow fiber membranes, where it served to effectively debitter grapefruit juice, improving the sensory properties to an acceptable level of bitterness (Gray and Olsen, 1981). More recently, it has been reported that naringinase, immobilized in cellulose acetate in a bioreactor (Tsen, 1990) or in packaging films just prior to casting (Soares and Hotchkiss, 1998), could actively decrease the naringin content of grapefruit juice.

6.1.4.2 Adsorbents   Resins and adsorbents were considered for debittering citrus juices and isolation of flavonoids when the importance of these compounds was recognized. In Australia, navel orange juice bitterness reduction has long been a research priority, and adsorbent resin studies have achieved some degree of commercialization (Johnson and Chandler, 1989). The ability of styrene–divinylbenzene resins to reduce bitterness from limonin and naringin during contact with citrus juices has been reported (Puri, 1984). Some properties and conditions for application of certain of these resins to juice debittering have been reported (Manlan et al., 1990; Matthews et al., 1990).

Although it may not have the capacity of some adsorbents, β-cyclodextrin has an ability to form inclusion complexes with limonin and naringin, thereby reducing juice bitterness (Konno et al., 1982). Simultaneous reduction of juice bitterness and acidity without flavor loss has also been achieved using ion exchange and adsorbent resins (Mitchell et al., 1985). The mineral adsorbent, Florisil (magnesium silicate) is also capable of reducing both bitterness and acidity of grapefruit juice, as demonstrated by taste panel studies (Barmore et al., 1986). Commercial application of anionic exchange resins for acid reduction has resulted in the FDA-approved product, reduced-acid frozen concentrated orange juice (FDA, 1992a). Kinetics of limonin and naringin removal and reduction of titratable acid from juices by resins and adsorbents have been studied, with recommendations for effectiveness (Johnson and Chandler, 1989).

Commercial effectiveness of adsorbents for debittering navel orange and grapefruit juices without significantly altering juice properties or other chemical components has been demonstrated (Norman, 1990; Kimball and Norman, 1990). Divinylbenzene polymers have been approved for juice bitterness reduction, if proper washing occurs prior to use (FDA, 1992b). These processes are effective and generally conform to maintaining the standards of identity of the juice to which it is applied. Process steps first require juice pasteurization, followed by membrane filtration to produce a clarified permeate and pulpy retentate. The permeate is passed through a column containing the adsorbent to lower the concentration of the bitter agents, and then recombined with the retentate to make the finished product. At the end of a process cycle, the system is flushed with water, cleaned with a caustic solution, and chemically sanitized (Wethern, 1991).

Special care must be taken to assure the debittering process is sanitary, otherwise the resin bed can plug or ferment from microbial growth. The process is costly, as membranes and resins have limited lifetimes and the technology requires some sophistication for proper operation. The added process cost to the value of the juice must also be considered. Navel orange and grapefruit juices have lower economic value because they do not have large market appeal; thus, the quality improvement from debittering may be questionable, especially when bitter products may be improved by simply blending with very late season juice.

6.2 PRODUCT STABILITY AND PACKAGING

There are many chemical aspects relating processing conditions, product stability, and container type to citrus juice quality; thus, the following discussion will be limited to only a few practical examples. Mentioned in the opening paragraph of this chapter, these include the effects of thermal treatment, oxygen, and package type on juice nutrient retention, flavor changes, and browning. The preservative effect of cold temperature on juice flavor quality and vitamin C loss by oxidation has been recognized since the 1930s; however, even frozen juice must be deaerated to attain a significant shelf life (Shrader and Johnson, 1934; Eddy, 1936).

6.2.1 Vitamin C Loss

Citrus juice, a very complex, highly flavored juice, is more sensitive than, for example, apple juice to degradative conditions. Although there are realistic boundaries in actual processing environments, the higher the temperature and shorter the time, the better the juice quality. The boundaries apply to the upper temperature limit resulting in extreme chemical degradation and the difficult prospect of trying to control a thermal process (e.g., a juice heat exchanger) time of less than a couple seconds. If one wishes to consider the effect of processing at two different temperatures on the loss of vitamin C, a common and defensible chemical approach is to apply the Arrhenius equation (Toledo, 1991) in its linear form:

$$\ln k = (-E_a/R)(1/T_2 - 1/T_1)$$

where k is the degradation at higher relative to lower temperature, E_a is the activation energy, measured in cal/mol, R is the gas constant, 1.987 cal/mol K, and T_1, T_2 are the lower and higher temperatures in K (K = °C + 273).

To apply this equation to a citrus juice situation, suppose one wanted to determine the effect on vitamin C by comparing juice heated in a pasteurizer at 75°C for 30 sec with 90°C for 5 sec. The E_a for vitamin C is approximately 23 kcal/mol. Substituting in the above equation and solving for the rate constant:

$$\ln k = (-23,000/1.987)(1/363 - 1/348)$$

$$\ln k = 1.43 \qquad k = 4.2$$

The result implies that vitamin C degradation is 4.2 times faster at 90°C than at 75°C, or that 1 sec at 90°C has vitamin C loss equivalent to 4.2 sec at 75°C. In commercial practice, it is realistic to compare a time of 30 sec at 75°C with 5 sec at 90°C. The result may be unexpected, finding there is 30/(4.2 × 5), or approximately 1.4 times more degradation of vitamin C with the 75°C process. This finding generally proves that higher temperature, short time processes result in citrus juice products with higher quality.

Vitamin C loss during the evaporation process for a typical citrus juice evaporator has also been studied using the above method. It can be determined that although flashing the juice into the vacuum at the beginning of the process lowers the oxygen content to less than 0.5 ppm, oxidative vitamin C losses were many times greater than anaerobic losses (Braddock and Sadler, 1989). Considering losses during the evaporation process, even if no oxygen was removed and all was devoted to vitamin C degradation in the juice, the total reduction would only be 4–5 mg of the 40–60 mg/100 mL initially present. The practical importance of the two situations discussed here is that loss of vitamin C during either juice pasteurization or concentration may not be as significant as losses during storage after packaging.

6.2.2 Flavor Deterioration

Although disappointing, as of this writing, there has been no technological breakthrough or identification of flavor constituents, which would enable thermally pasteurized juice or reconstituted concentrate to taste like fresh juice. Chemical deterioration of juice flavor during thermal processing is more complex than loss of a chemical compound, such as vitamin C. Many research articles have been published about composition and analysis and the effects of process variables on the volatile flavor components of juice. They will not be discussed in this text, except to state that heating irreversibly and negatively alters juice flavor, so that it no longer has the aroma and character of fresh juice. Evaporation is known to cause flavor loss, and, combined with membrane processing, additional losses may have negative impacts on the implementation of such combined processes (Johnson et al., 1996).

Some process developments have resulted in improved flavor because they minimize application of heat. Notable processes freeze concentration, reverse osmosis, and high-pressure pasteurization have been commercialized but are not commonly applied in the industry (Braddock, 1986; Braddock and Marcy, 1987; Braddock and Sadler, 1989; Braddock et al., 1998). Capital investment costs are high, freeze concentration and reverse osmosis cannot feasibly achieve 65 °Brix, and high-pressure pasteurization is limited by being a batch or semicontinuous process. Another process, pulsed electric field pasteurization is being studied but does not yet have commercial citrus juice application because the advantages and mechanism of the process have not been completely defined (Vega-Mercado et al., 1997; Mermelstein, 1998). Microwave juice pasteurization has been studied (Nikdel et al., 1993), but probably has little flavor advantage, because the juice still must be heated. Also, a separate heat exchanger must be used to cool the hot juice, negating any presumed advantage of not using a steam heat exchanger to heat the juice.

Considering citrus juice flavor deterioration from thermal processes taste panelists can distinguish differences between products heated at extreme temperatures (111°C) with lower temperatures (80–97°C) but not between the lower temperatures (Braddock and Marcy, 1987). The explanation for this lies in the complexity of the flavor, as well as in the fact that most chemicals in juices have low activation energies for degradation (under 50 kcal/mol). Comparatively, destruction of microorganisms and pectinesterase (E_a>50 kcal/mol) requires thermal processes at conditions greater than that which causes chemical change. It follows that reaction of flavor compounds with oxygen during packaging and storage, like for vitamin C, may be as significant for product quality as effects of process temperature.

6.2.3 Browning

Like flavor deteriorating, much has been written about browning of citrus juices. In general, browning of citrus juices is a chemical result of the non-oxidative degradation of vitamin C (ascorbic acid) in the presence of reducing sugars and amino acids at the juice pH. The actual chemical reaction scheme is not thoroughly defined, but absence of any of these components will not prevent browning. Severe process times and temperatures can cause or initiate browning; however, they may not be as significant as the finished product storage environment. Rate constants for vitamin C degradation, as well as browning, in a clarified orange juice serum have been determined relative to effects of storage temperature and soluble solids concentration in an anaerobic environment (Johnson et al., 1995). Researchers have determined that furfuraldehyde is a reaction product of ascorbic acid degradation in the browning pathway. This and other volatile carbonyls are usually detected in a product before the browning becomes apparent. Browning resulting from oxidation reactions has not previously been considered because most juices were packaged in oxygen-impermeable glass or cans or chilled to slow reaction rates and deterioration processes.

6.2.4 Packaging

Juice packaged in glass or metal containers can undergo oxidative reactions only to the extent of the amount of oxygen in the headspace after closure. The importance of deaeration before packaging these juices has received much debate but may not be too significant for product kept chilled during a consumer product lifetime of only several months. If the product is packaged aseptically in glass for ambient temperature storage, as in Europe, deaeration may extend the shelf life somewhat. This is difficult to prove because nonoxidative deterioration reactions occur concurrently and rapidly at ambient temperature. Another factor to be considered is that European juice is manufactured mostly from reconstituted concentrate, with the disadvantage to initial flavor quality of an additional heat treatment, after the heat of concentration at the point of manufacture. A potential problem related to microbial stability has also been identified with the identification of molds present in the paperboard part of cartons used for packaging citrus juices (Narciso and Parish, 1997).

Use of polymeric containers for citrus juices has the big advantage of cheaper packages. Aseptic processes for juices in polymeric packages are also common but suffer a few major disadvantages compared to glass. Polymer containers may allow permeation of oxygen into the product, either by diffusion through the polymer or from leaks at the seals or edges. This permeation is also rate dependent on the temperature, complicating the degradative processes. Chemical reactions causing loss of flavor and nutrients and increased browning by oxidation then shorten the shelf stability considerably, to less than a few months. Generally, package oxygen permeability barriers are not correlated with the flavor barrier properties. One must also be aware that low activation energies for degradation reactions dictate that storage stability challenge tests above 40°C have little chance to provide results useful for ambient temperature shelf life predictions. This is because chemical reactions above this point are unlikely to occur at ambient conditions.

Flavor absorption (scalping), another problem associated with polymer packages, may be dealt with by adding extra flavor to products kept chilled. This may have limited effectiveness for aseptic products at ambient temperature because the nonpolar polymer (e.g., polyethylene) has strong affinity for terpenes and flavor molecules, rapidly (within a few hours) decreasing their concentration in the juice (Durr et al., 1981; Sadler and Braddock, 1991). Reactions of scalped terpene compounds with oxygen within the polymer matrix may also allow off-flavors of these molecules the chance to diffuse back into the juice (Kutty et al., 1994). There is also evidence that scalping may not be too important to maintaining the sensory quality of orange juice (Pieper et al., 1992), although this may be open to discussion.

A good discussion of the importance of packaging to quality of many foods has been published (Blumenthal, 1997). Both the flavor system and its interaction with the package material must be considered. Important flavor system factors are product usage level, cost, and stability (reactivity, volatility). Flavor change, loss of product acceptability, changes of barrier properties, and package failure (delamination, leaks) may result from package-flavor interaction. The above discussion on incorporating enzymes into package films also has potential for altering or stabilizing packaged products. This subject, called "active packaging," offers some future applications for citrus juices. However, knowledge of chemical degradation kinetics and study of many variables leads one to the conclusion that the best thing to do to preserve packaged citrus juice quality is to keep it cold (Graumlich et al., 1986; Sizer, et al., 1988).

6.3 NONTHERMAL PROCESSING

Current inactivation of pectinesterase is accomplished by heat treatment (Chapter 5); however, nonthermal procedures to inactivate microorganisms such as pulsed electric fields (PEF) (Vega-Mercado et al., 1997) and high-pressure processing may also affect the activity of this enzyme in citrus juice. The PEF method uses high-intensity electric field discharges to kill microorganisms by some poorly understood mechanism of pore formation in cellular membranes. As heat is also generated, heat exchangers may be required in flow schemes for liquid food applications with recycling for adequate kill. Application of PEF to inactivate pectinesterase in citrus juices has not been adequately demonstrated.

High-pressure processing (HPP) of citrus juices and other foods has been commercialized in a semi-continuous process (Flow, 1998). In batch processing, the raw product is poured into polymeric packaging, placed in the HPP

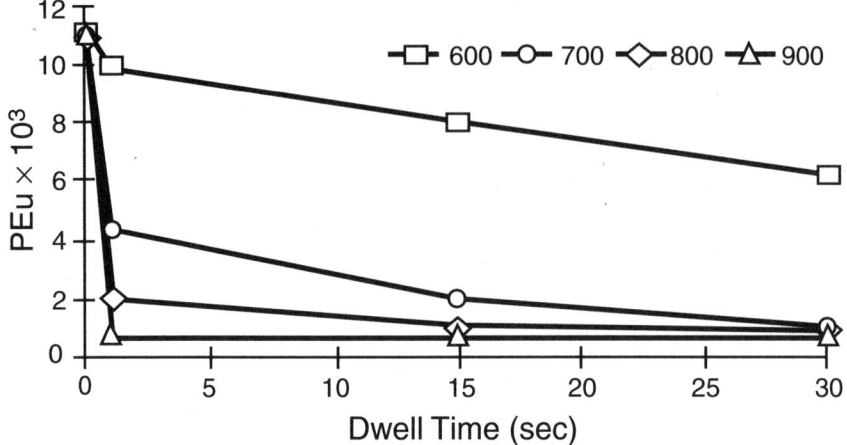

Figure 6.3 High-pressure processing inactivation of orange juice pectinesterase at 600, 700, 800, and 900 MPa for three dwell times. Dwell time is the time the sample remains at the set pressure (Goodner et al., 1998).

unit, and treated. Repackaging the product is not required. This technique has been demonstrated to kill microorganisms and effectively inactivate pectinesterase, with no significant flavor change from fresh juice. The HPP method at pressures greater than 700 MPa were effective in inactivating the heat-labile form of pectinesterase in orange juice (Fig. 6.3) but showed no effect on the heat-stable form of this enzyme (Goodner et al., 1998). This study also showed that if treatment times were for at least 1 min at pressures above 700 MPa, the

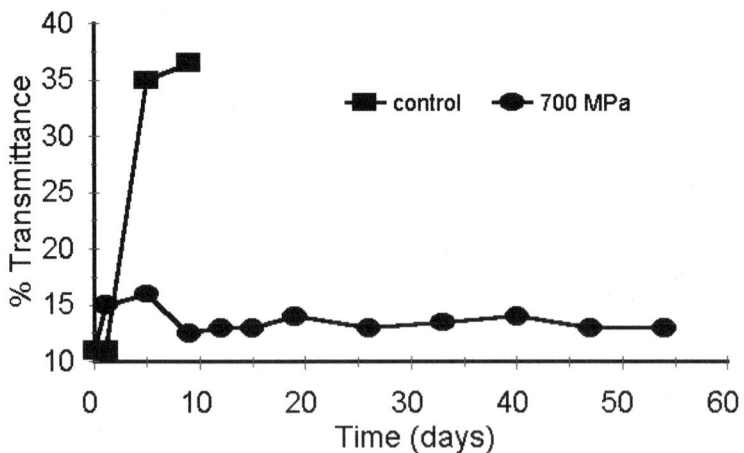

Figure 6.4 Cloud stability of orange juice pressurized for 1 min at 700 MPa (Goodner et al., 1998).

juice cloud was stable for over 2 months with no growth of microorganisms (Fig. 6.4). Much lower pressures (<100 MPa) using CO_2 to lower juice pH with carbonic acid, will partially inactivate pectinesterase and confer juice cloud stability (Balaban et al., 1995; Chapter 5).

REFERENCES

Albach, R. F., Redman, G. H., Cruse, R. R., and Petersen, H. D. (1981). Seasonal variation of bitterness components, pulp, and vitamin C in Texas commercial citrus juices, *J. Agric. Food Chem.* **29**:805–808.

Attaway, J. A. and Carter, R. D. (1975). Symposium—Analytical indicators of processed orange juice quality, 1972–73 and 1973–74, *Proc. FL. State Hort. Soc.* **88**: 339–370.

Baker, R. A. and Cameron, R. G. (1999). Clouds of citrus juices and juice drinks, *Food Technol.* **53**(1):64–69.

Balaban, M. O., Marshall, M. R., and Wicker, L. (1995). Inactivation of enzymes in foods with pressurized CO_2. U.S. Patent 5,393,547.

Balls, A. K. (1949). Enzyme problems in the citrus industry, *Food Technol.* **3**(3):96–100.

Ben-Shalom, N. and Pinto, R. (1986). The role of hesperidin in the turbidity of orange albedo aqueous extract, *Lebensm.-Wiss. u.-Technol.* **19**:158–160.

Barmore, C. R., Fisher, J. F., Fellers, P. J. and Rouseff, R. L. (1986). Reduction of bitterness and tartness in grapefruit juice with Florisil®. *J. Food Sci.* **51**(2):415–416, 439.

Blumenthal, M. M. (1997). How food packaging affects food flavor, *Food Technol.* **51**(1):71–74.

Braddock, R. J. (1986). Quality of freeze concentrated grapefruit juice, *Proc. Int. Fruchtsaft-Union Sympos*, XIX. Den Haag, The Netherlands, pp. 89–95.

Braddock, R. J. and Marcy, J. E. (1987). Quality of freeze concentrated orange juice, *J. Food Sci.* **52**:159–162.

Braddock, R. J. and Sadler, G. D. (1989). "Chemical changes in citrus juices during concentration processes." In *Quality Factors of Fruits and Vegetables*: *Chemistry and Technology*, J. J. Jen, Ed., ACS Symposium Series No. 405, Washington, D.C., Chap. 22, pp. 293–304.

Braddock, R. J., Parish, M. E., and Goodner, J. K. (1998). High pressure pasteurization of citrus juices, *Trans. Citrus Eng. Conf.* **44**:1–10.

Cameron, R. G., Baker, R. A., and Grohmann, K. (1997). Citrus tissue extracts affect juice cloud stability, *J. Food Sci.* **62**(2):242–245.

Cameron, R. G., Baker, R. A., and Grohmann, K. (1998). Multiple forms of pectinmethylesterase from citrus peel and their effects on juice cloud stability, *J. Food Sci.* **63**(2):253–256.

Chandler, B. V. and Robertson, G. L. (1983). The solubility of limonin, the bitter principle of orange juice, *J. Sci. Food Agric.* **84**:1272–1284.

Durr, P., Schobinger, U., and Waldrogel, R. (1981). Aroma quality of orange juice after filling and storage in soft packages and glass bottles, *Lebensm.-Verpackung.* **20**:91–93.

Drewnowski, A., Henderson, S. A., and Shore, A. (1997). Taste responses to naringin, a flavonoid, and the acceptance of grapefruit juice are related to genetic sensitivity to 6-*n*-propylthiouracil, *Am. J. Clin. Nutr.* **66**:391–397.

Dunlap, W. J., Hagen, R. E., and Wender, S. H. (1962). Preparation and properties of rhamnosidase and glucosidase fractions from a fungal flavonoid glycosidase preparation, "Naringinase C-100," *J. Food Sci.* **27**:597–601.

Eddy, C. W. (1936). Absorption rate of oxygen by orange juice, *Ind. Eng. Chem.* **28**(4): 480–483.

FDA (1992a). U.S. Food and Drug Administration, HHS, 21 CFR Ch. 1, 146.148.

FDA (1992b). U.S. Food and Drug Administration, HHS, 21 CFR Ch. 1, 173.65.

Fellers, P. J., de Jager, G., Poole, M. J., Hill, E. C., and Mittal, P. (1986). Quality of Florida-packed retail grapefruit juices as determined by consumer sensory panels and chemical and physical analyses, *J. Food Sci.* **51**(2):417–420.

Flow International (1998). Fresher under pressure. Brochure. Flow International Corp., Kent, Washington, USA and Washington, D.C., USA. www.flowcorp.com.

Goodner, J. K., Braddock, R. J., and Parish, M. E. (1998). Inactivation of pectinesterase in orange and grapefruit juices by high pressure, *J. Agric. Food Chem.* **46**(5):1997–2000.

Grant, P. M. (1989). Citrus juice concentrate processor, U.S. Pat. 4,886,574.

Graumlich, T. R., Marcy, J. E., and Adams, J. P. (1986). Aseptically packaged orange juice and concentrate: A review of the influence of processing and packaging conditions on quality, *J. Agr. Food Chem.* **34**:402–405.

Gray, G. M. and Olson, A. C. (1981). Hydrolysis of high levels of naringin in grapefruit juice using a hollow fiber naringinase reactor, *J. Agric. Food Chem.* **29**:1298–1301.

Guadagni, D. G., Maier, V. P., and Turnbaugh, J. G. (1974). Some factors affecting sensory thresholds and relative bitterness of limonin and naringin, *J. Sci. Fd. Agric.* **25**:1199–1205.

Hasegawa, S. and Maier, V. P. (1983). Solutions to the limonin bitterness problem of citrus juices. *Food Technol.* **37**(6):73–77.

Hasegawa, S., Brewster, L. C., and Maier, V. P. (1975). Limonoate:NAD(P) oxidoreductase and debittering of citrus juices and other products, U.S. Pat. 3,920,851.

Hasegawa, S., Bennett, R. D., Herman, Z., Fong, C. H. and Ou, P. (1989). Limonoid glucosides in citrus, *Phytochemistry* **28**(6):1717–1720.

Hasegawa, S., Vandercook, C. E., Choi, G. Y., Herman, Z., and Ou, P. (1985). Limonoid debittering of citrus juice sera by immobilized cells of *Corynebacterium fascians*, *J. Food Sci.* **50**:330–332.

Herman, Z., Hasegawa, S., and Ou, P. (1985). Nomilin acetyl-lyase, a bacterial enzyme for nomilin debittering of citrus juices, *J. Food Sci.* **50**:118–120, 124.

Horowitz, R. M. and Gentili, B. (1977). "Flavonoid constituents of citrus." In *Citrus Science and Technology*, Vol. 1, S. Nagy, P. E. Shaw, and M. K. Veldhuis, Eds., AVI, Westport, CT.

Hsu, W. J., Berhow, M., Robertson, G. H., and Hasegawa, S. (1998). Limonoids and flavonoids in juices of Oroblanco and Melogold grapefruit hybrids, *J. Food Sci.* **63**(1):57–60.

Jansen, E. F., Jang, R., and Bonner, J. (1960). Orange pectinesterase binding and activity, *Food Res.* **25**:64–72.

Johnson, R. L. and Chandler, B. V. (1989). Kinetic studies of adsorption of bitter principles and titratable acid from grapefruit juice, *J. Chem. Tech. Biotechnol.* **44**: 225–236.

Johnson, J. R., Braddock, R. J., and Chen, C. S. (1995). Kinetics of ascorbic acid loss and nonenzymatic browning in orange juice serum: Experimental rate constants, *J. Food Sci.* **60**(3):502–505.

Johnson, J. R., Braddock, R. J., and Chen, C. S. (1996). Flavor losses in orange juice during ultrafiltration and subsequent evaporation, *J. Food Sci.* **61**(3):540–543.

Jourdan, P. S., Mansell, R. L., Oliver, D. G., and Weiler, E. W. (1984). Competitive solid phase enzyme-linked immunoassay for the quantification of limonin in citrus, *Anal. Biochem.* **138**:19–24.

Kimball, D. A. and Norman, S. I. (1990). Changes in California Navel orange juice during commercial debittering, *J. Food Sci.* **55**(1):273–274.

Klavons, J. A., Bennett, R. D., and Vannier, S. H. (1991). Nature of the protein constituent of commercial orange juice cloud, *J. Agr. Food Chem.* **39**(9):1545–1548.

Konno, A., Misaki, M., Toda, J., Wada, T., and Yasumatsu, K. (1982). Bitterness reduction of naringin and limonin by β-cyclodextrin, *Agric. Biol. Chem.* **46**(9):2203–2208.

Krop, J. J. P. and Pilnik, W. (1974). Effect of pectic acid and bivalent cations on cloud loss of citrus juice, *Lebensm.-Wiss. u. Technol.* **7**(1):62–63.

Kutty, V., Braddock, R. J., and Sadler, G. D. (1994). Oxidation of *d*-limonene in presence of low density polyethylene, *J. Food Sci.* **59**(2):402–405.

Manlan, M., Matthews, R. F., Rouseff, R. L., Littell, R. C., Marshall, M. R., Moye, H. A., and Teixeira, A. A. (1990). Evaluation of the properties of polystyrene divinylbenzene adsorbents for debittering grapefruit juice, *J. Food Sci.* **55**(2):440–445, 449.

Matthews, R. F., Rouseff, R. L., Manlan, M., and Norman, S. I. (1990). Removal of limonin and naringin from citrus juice by styrene-divinylbenzene resins, *Food Technol.* **44**(4):130–132.

Maier, V. P., Brewster, L. C., and Hsu, A. C. (1971). Development of methods for producing nonbitter Navel orange juice, *Citrograph.* **56**(Sept.):373–375.

Mermelstein, N. H. (1998). Interest in pulsed electric field processing increases, *Food Technol.* **52**(1):81–82.

Mitchell, D. H., Pearce, R. M., Smith, C. B., and Brown, S. T. (1985). Removal of bitter naringin and limonin from citrus juices containing the same, U.S. Pat. 4,514,427.

Narciso, J. A. and Parish, M. E. (1997). Endogenous microflora of gable-top carton paperboard used for packaging fruit juice, *J. Food Sci.* **62**(6):1223–1225, 1239.

Nicol, K. J. and Chandler, B. V. (1978). The extraction of the enzyme degrading the limonin precursor in citrus albedo, *J. Sci. Fd. Agric.* **29**:795–802.

Nikdel, S., Chen, C. S., Parish, M. E., MacKellar, D. G., and Friedrich, L. M. (1993). Pasteurization of citrus juice with microwave energy in a continuous-flow unit, *J. Agric. Food Chem.* **41**(11):2116–2119.

Norman, S. I. (1990). A commercial citrus debittering system, *Trans. Citrus Eng. Conf.* **36**:1–31.

Ono, M., Tosa, T., and Chibata, I. (1977). Preparation and properties of naringinase immobilized by ionic binding to DEAE Sephadex®, *J. Ferment. Technol.* **55**:493–500.

Pieper, G., Borgudd, L., Ackermann, P., and Fellers, P. (1992). Absorption of aroma volatiles of orange juice into laminated carton packages did not affect sensory quality, *J. Food Sci.* **57**(6):1408–1411.

Poore, H. D. (1934). Recovery of naringin and pectin from grapefruit residue, *Ind. Eng. Chem.* **26**(6):637–639.

Pulley, G. N. (1936). Solubility of naringin in water, *Ind. Eng. Chem.* **8**(5):360.

Puri, A. (1984). Preparation of citrus juices, concentrates and dried powders which are reduced in bitterness, U.S. Pat. 4,439,458.

Puri, M., Marhawa, S. S., and Kothari, R. M. (1996). Studies on the applicability of alginate entrapped naringinase for the debittering of Kinnow juice, *Enzyme Microbiol. Technol.* **18**(4):281–285.

Ram, B. P., Jang, L., Martins, L., and Singh, P. (1988). An improved enzyme immunoassay for limonin, *J. Food Sci.* **53**(1):311–312.

Ranganna, S., Govindarajan, V. S., and Ramana, K. V. R. (1983a). Citrus fruits — Varieties, chemistry, technology, and quality evaluation, Part II. Chemistry, technology, and quality evaluation. A. Chemistry, *CRC Crit. Rev. Food Sci. Nutr.* **18**(4): 313–386.

Ranganna, S., Govindarajan, V. S., and Ramana, K. V. R. (1983b). Citrus fruits. Part II. Chemistry, technology, and quality evaluation. B. Technology, *CRC Crit. Rev. Food Sci. Nutr.* **19**(1):1–98.

Redd, J. B., Hendrix, D. L., and Hendrix, C. M., Jr. (1986). *Quality Control Manual for Citrus Processing Plants*, Vol I, Intercit, Safety Harbor, FL.

Rombouts, F. M. Versteeg, C., Karman, A. H., and Pilnik, W. (1982). "Pectinesterases in components of citrus fruits related to problems of cloud loss and gelation in citrus products." In Use of Enzymes in Food Technology. Symp. International, P. Dupuy, Ed., Versailles, May 5–7, pp. 483–487.

Rothschild, G. and Karsenty, A. (1974). Cloud loss during storage of pasteurized citrus juices and concentrates, *J. Food Sci.* **39**:1037–1041.

Rouse, A. H. (1953). Distribution of pectinesterase and total pectin in component parts of citrus fruits, *Food Technol.* **7**(9):360–362.

Rouse, A. H., Atkins, C. D., and Moore, E. L. (1962). Seasonal changes occurring in the pectinesterase activity and pectic constituents of the component parts of citrus fruits. I. Valencia oranges, *J. Food Sci.* **27**(5):419–425.

Rouse, A. H., Atkins, C. D., and Moore, E. L. (1965). Seasonal changes occurring in the pectinesterase activity and pectic constituents of the component parts of citrus fruits. III. Silver Cluster Grapefruit, *Food Technol.* **19**(4):241–244.

Rouseff, R. L. and Matthews, R. F. (1984). Nomilin, taste threshold and relative bitterness, *J. Food Sci.* **49**(3):777–779, 790.

Rouseff, R. L., Martin, S. F., and Youtsey, C. O. (1987). Quantitative survey of narirutin, naringin, hesperidin and neohesperidin in citrus, *J. Agr. Food Chem.* **35**(6):1027–1030.

RSK-Values (1987). *The Complete Manual*, VdF Flüssiges Obst Gmbh, D-5429 Schönborn, Germany.

Sadler, G. D. and Braddock, R. J. (1991). Absorption of citrus flavor volatiles by low density polyethylene, *J. Food Sci.* **56**(1):35–37, 54.

Scott, W. C., Kew, T. J., and Veldhuis, M. K. (1965). Composition of orange juice cloud, *J. Food Sci.* **30**:833–837.

Shaw, P. E. and Wilson, C. W. (1988). Quantitative determination of limonin in citrus juices by HPLC using computerized solvent optimization, *J. Chrom. Sci.* **26**(Sept): 478–481.

Shaw, P. E., Calkins, C. O., McDonald, R. E., Greany, P. D., Webb, J. C., Nisperos-Carriedo, M. O., and Barros, S. M. (1991). Changes in limonin and naringin levels in grapefruit albedo with maturity and the effects of gibberellic acid on these changes, *Phytochemistry* **30**(10):3215–3219.

Shrader, J. H. and Johnson, A. H. (1934). Freezing orange juice, *Ind. Eng. Chem.* **26**(8): 869–874.

Sizer, C. E., Waugh, P. L., Edstam, S., and Ackermann, P. (1988). Maintaining flavor and nutrient quality of aseptic orange juice, *Food Technol.* **42**(6):152–159.

Soares, N. F. F. and Hotchkiss, J. H. (1998). Naringinase immobilization in packaging films for reducing naringin concentration in grapefruit juice, *J. Food Sci.* **63**(1):61–65.

Termote, F., Rombouts, F. M., and Pilnik, W. (1977). Stabilization of cloud in pectinesterase active orange juice by pectic acid hydrolysates, *J. Food Biochem.* **1**:15–34.

Thomas, D. W., Smythe, C. V., and Labbee, M. D. (1958). Enzymatic hydrolysis of naringin, the bitter principle of grapefruit, *Food Res.* **23**:591–598.

Ting, S. V. (1958). Enzymatic hydrolysis of naringin in grapefruit, *J. Agric. Food Chem.* **6**(7):546–549.

Toledo, R. T. (1991). "Kinetics of chemical reactions in foods." In *Fundamentals of Food Process Engineering*, 2nd ed., Van Nostrand Reinhold, New York, Chap. 8, pp. 310–314.

Tsen, H. Y. (1990). Immobilized *Penicillium* sp. *naringinase* and its use in removing naringin and limonin from fruit juice, U.S. Pat. 4,971,812.

Vega-Mercado, H., Martin-Belloso, O., Qin, B. L., Chang, F. J., Góngora-Nieto, M. M., Barbosa-Cánovas, G. V., and Swanson, B. G. (1997). Non-thermal food preservation: Pulsed electric fields, *Trends Food Sci. Technol.* **8**:151–157.

Versteeg, C., Martens, L. J. H., Rombouts, F. M., Voragen, A. G. J., and Pilnik, W. (1977). Enzymatic hydrolysis of naringin in grapefruit juice, *Lebensm.-Wiss. u. Technol.* **10**(5):268–272.

Versteeg, C., Rombouts, F. M., Spaansen, C. H., and Pilnik, W. (1980). Thermostability and orange juice cloud destabilizing properties of multiple pectinesterases from orange, *J. Food Sci.* **45**:969–971, 998.

Vitali, A. A. and Rao, M. A. (1984). Flow properties of low-pulp concentrated orange juice: Serum viscosity and effect of pulp content, *J. Food Sci.* **49**(3):876–881.

Wethern, M. (1991). Citrus debittering with ultrafiltration/adsorption combined technology, *Trans. Citrus Eng. Conf.* **37**:48–66.

CHAPTER 7

PULP WASHING, JUICE PULP RECOVERY, AND UTILIZATION

During the juice recovery processes described in Chapter 4, the pulp and pulpy juice streams may be a source of two by-products, which may be recovered in a separate process operation for specific applications in juices and food products. This pulp contains juice soluble solids not completely expressed in the finishing operation that may be recovered by water extraction. The juice and soluble solids extracted in this manner are commonly called pulp wash. In Florida, there are regulations defining the product and its use as water-extracted soluble fruit solids (WESFS), or commonly, washed pulp solids (Florida Department of Citrus, 1998). Pulp wash has a controversial history in Florida due to quality issues and growers' concerns that some users might prefer to purchase cheaper pulp wash concentrate instead of juice concentrate.

7.1 PULP WASHING PROCESS

Attempts to recover additional juice from the fruit during times when fruit availability was low (such as after freezes) resulted in development of the pulp washing process. Early studies recognized that juice yields could be improved by recovering additional juice solids from the primary finisher pulp by water extraction. The amount of juice remaining in finisher pulp was found to vary with process parameters by a method for measuring the titratable acidity of hot water extracts of pulp samples (Kilburn and Dillman, 1951). A description of the process and properties of the product were described in a preliminary study (Olsen et al., 1958). For the process to be economical, pilot-scale studies showed that higher efficiencies could be achieved if the water rinsing was performed in a multistaged countercurrent extraction process (Wenzel et al., 1959).

A generic illustration of a typical four-stage pulp washing process with mixers, finishers, and product flows is presented in Figure 7.1. Each stage consists of a mixing device, which may be a mixing screw or tank, and a finisher. The pulp put into the process at stage 1 exits at stage 4 as spent pulp, washed free of most soluble solids to be in the range of <1 °Brix. Pure water enters at stage 4 and exits at stage 1 as the product pulp wash or strong liquor. Soluble solids concentration of the strong liquor is variable, determined by process parameters, but is commonly in the range of 4–7 °Brix for oranges. The spent pulp may be sent to the feed mill or have by-product uses, and the pulp wash may be concentrated as a product similar to juice concentrate or may be combined with the juice prior to evaporation. Following the severe 1962 freeze in Florida, commercial interest in pulp washing to increase soluble solids yield expanded rapidly. The major processing equipment manufacturers helped develop processes and provide more refined information about them to their customers.

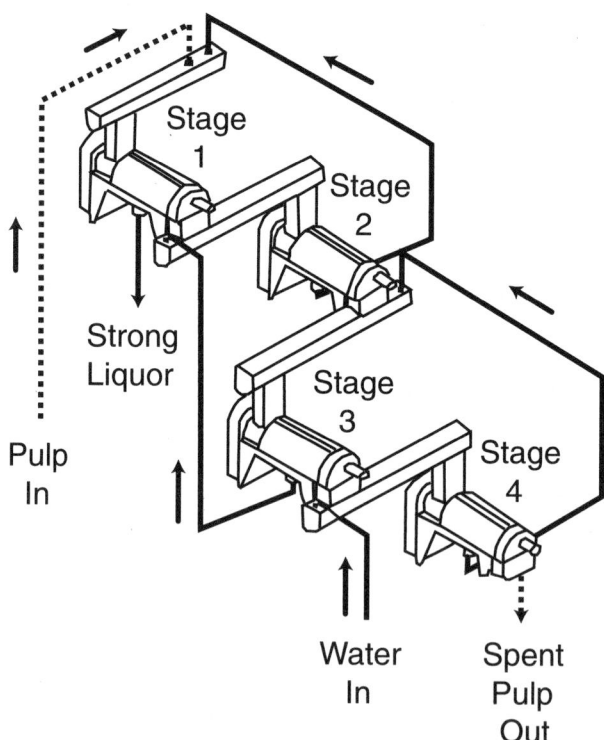

Figure 7.1   Four-stage countercurrent pulp washing system illustrating pulp and pulp wash liquor flows. Each stage consists of a mixer and finisher.

7.1.1 FMC Process

7.1.1.1 Pulp Wash   Washing pulp from the FMC juice extraction process (FMC FoodTech, Citrus Systems Division, Lakeland, FL) was described in a study comparing four-stage and seven-stage systems (Belk, 1964). This report documented the importance of key process variables, equilibrium mixing of water and pulp, control of flow rates, and system sanitation to provide uniform operation. Mass flow data and algebraic calculations showed that at water/pulp ratios from 1/1 to 1.5/1, seven-stage were more efficient than four-stage systems, requiring less evaporation. Using mass values of 8 lb pulp/box and 1.5/1 water/pulp ratio, wash liquor sent to the evaporator was 7.4 °Brix for the seven-stage and 6.9 °Brix for the four-stage system. However, additional finishers, mixers, and installation were required for the seven-stage system, partly off-setting the advantage of more efficient extraction. Recently, a pulp wash system utilizing static mixers in place of the traditional mixing screws or paddles has been developed. This system (Fig. 7.2) is more compact, safer, sanitary, and has less moving parts than a traditional system. Design features include the static mixers and finishers of each stage as well as vertical cylindrical tubes for water addition and conveying of the raffinate (pulp/water).

The yield of soluble solids is intimately associated with the quantity of pulp sent to the washing process. A characteristic of the FMC juice extraction process is that preliminary finishing, or separation of juice pulp from rag, core, membrane, and the two peel plugs cut from each fruit, occurs in the strainer tube of the juice extractor. Thus, pulp from the primary juice finisher entering the washing process is mostly composed of ruptured juice vesicles. In the FMC process, pulp from juice finishers sent to be washed may range in amount from about 3 to 8 lb/box for oranges at commercial extractor settings (Ballentine and Ferguson, 1985). The variables that determine pulp yield are primarily

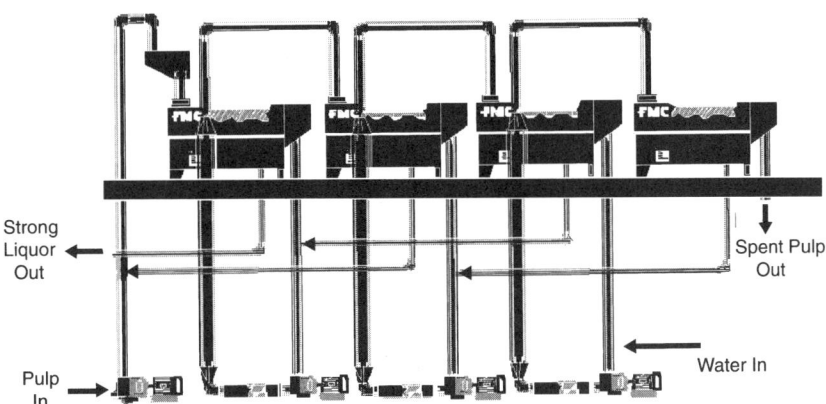

Figure 7.2   FMC continuous static mixer, four-stage pulp washing system. (By permission of FMC FoodTech, Citrus Systems Division, Lakeland, FL.)

related to the juice extractor setup, finisher setting, and condition of the fruit. For example, maximum yield of pulp for washing might range from 8 to 27 lb/box with conditions of hard squeeze, large strainer tube hole diameter, small diameter orifice discharge, and a loose finish (Ballentine and Ferguson, 1985). Keep in mind that the pulp wash is the secondary product, which must be balanced both economically and in quality, with the primary product, the juice. Because of this, juice takes precedence over pulp wash recovery in process design and efficiency.

7.1.1.2 Core Wash In some situations, to obtain additional solids, processors with FMC extractors recover and wash the core material ejected from the orifice discharge beneath the juice strainer tube of the extractors. The core material includes the rag membrane, core, seeds, and the two peel plugs compressed in the strainer tube during the extractors' juice recovery cycles. Core material recovery is accomplished by a small screw conveyor beneath the extractor line, separate from the larger conveyor, which takes the peel residue for the feed mill. When the core material is washed, only a single-stage washing with 1–2 parts water is commonly used because of bitterness and otherwise poor quality of the washed product (Kimball, 1991). Similar to pulp recovered from the juice finisher, the mass of the core material is dependent on the variables of the extraction process and fruit characteristics. Generally, the amount of core material from the extractors is in the range of 10–15 lb/box of oranges and soluble solids recovered may amount to about 0.2 lb/box (Table 7.1).

7.1.2 Brown Process

The development of a washing process for pulp from Brown juice extractors (Brown International Corporation, Covina, CA) was described by McKinnis et

TABLE 7.1 Amounts of Pulp and Soluble Solids from Orange and Grapefruit Pulp Wash Systems

	Pound Pulp/Box		Pound Soluble Solids/Box[a]	
	Valencia	Grapefruit	Valencia	Grapefruit
Brown pulp	4–5[b]	5–6[b]	0.5	0.4
Brown pomace	8[c]	11[b]	0.8	0.8
FMC pulp	4–5[d]	6[b]	0.5	0.5
FMC core	11[e]	—	0.2	—

[a]Pound soluble solids/box is calculated from pound/box based on 80% soluble solids recovery from a four-stage, 1.5/1 water/pulp process.
[b]Braddock and Kesterson (1976).
[c]Webb (1997).
[d]Ballentine and Ferguson (1985).
[e]Beasley (1998). Pound soluble solids/box estimated from single-stage system.

al. (1964). This study considered design features including equilibrium mixing, finisher variables, flow rate control, recoveries, and sanitation. Recovery of additional soluble solids by commercial processes may be in the range of 80–90% of the amount determined based on the dry weight of the extracted, or spent, pulp. The effect on the evaporator of adding pulp wash liquor to the juice stream reduced the °Brix equivalent to approximately 65% that of the juice stream, increasing the water removal cost. Increased amounts of water result in higher yields of soluble solids, but the pulp wash solids concentration will be lower, requiring more evaporation, thus more expense. In these processes, the soluble solids contained in the pulp feeding the washing steps is balanced by the solids in the final pulp wash out of the system and the solids remaining in the spent (washed) pulp.

The juice finishing process described in Chapter 4 for Brown International Corporation extractors determines the composition of the material sent to the pulp washing process. There are basically two pulp streams for washing: the pomace discharge from a primary finisher (scalper) having larger screen hole diameters and the pulp discharge from the secondary finisher. The scalper pomace contains membranous rag, seeds, and internal components from the juice stream, while ruptured juice vesicles and juice pass through the scalper screen to the secondary finisher, which discharges the pulp containing mostly juice vesicles. Washing systems may either wash the pomace from the scalping finisher, the pulp from the secondary finisher, or both streams combined. Estimates of the pulp and solids yields are presented in Table 7.1. It should be noted that when unwashed pulp is recovered as a by-product, most of the stream from the secondary finisher is unavailable for washing.

An engineering study of both cascade-type (Fig. 7.1) and pumped-flow pulp wash systems described a method for determining equilibrium conditions and maximum efficiency relating the important process variables of pulp and water flow, particle size, mixing and finishing conditions, and product °Brix (Webb, 1997). This procedure relates pulp and wash liquid °Brix measurement in system stages via equilibrium plots of modified McCabe–Thiele graphs. The plots generate equilibrium data and operating lines based on equations of the regression data of in-plant measurements for a number of commercial pulp wash systems. Use of this method should allow modeling of water addition, flow, and finisher operation for maximum operational efficiency of pulp wash systems and would be particularly useful as part of the design plan of new installations.

7.2 PULP WASH CONCENTRATE

The strong liquor from the pulp washing process (Fig. 7.1) may be sent to an evaporator for manufacture of the product, the pulp wash concentrate. The strong liquor is basically handled in the same manner as juice, coming from the first-stage finisher to a tank feeding a centrifuge for removal of extra sediment, then to a surge tank feeding the evaporator. A portion of the evaporator product may be recycled to raise the °Brix of the evaporator feed stream to near 10 °Brix, allowing more efficient evaporator operation. The final concentration for orange pulp wash evaporator pump-out may be near 65 °Brix, similar to juice concentrate. However, depending on specifications desired by the user, product may be between 50 and 65 °Brix.

7.2.1 Enzyme Use

7.2.1.1 *Viscosity Reduction* Operation of the evaporator during pulp wash concentration can be complicated by higher viscosity due to soluble pectin in the strong liquor. This problem may occur late in the season for certain fruit (e.g., grapefruits) or from extreme process variables during the washing steps. Since the problem is a result of too-high a pectin concentration in the strong liquor, the viscosity increases exponentially as the concentration increases in the various stages of the evaporator. The resultant concentrate may have viscosities of several thousand to as high as 20,000 mPa·s, where gels may form (Braddock and Kesterson, 1976). High viscosity in the evaporator lowers throughput, decreases energy efficiency, and can result in product with a cooked flavor.

Pectolytic enzymes are commonly added to the strong liquor to facilitate a considerable viscosity reduction, allowing trouble-free concentration to high °Brix in the evaporator. Depending on the enzyme used, concentrations to effect significant viscosity reduction range from 50 to 500 ppm of the soluble solids in the liquid. Holding times for this application are from 30 min to 1 hr prior to feeding the evaporator (Braddock and Kesterson, 1976). Application of enzymes for liquid viscosity reduction is a simple technique, which may be accomplished using two tanks, as shown in Figure 7.3. While one tank is feeding the evaporator, enzyme is reacting in the other. Optimization of this process for maximum cloud preservation would require the strong liquor to be pasteurized (90°C) immediately to inactivate the native pectinesterase. After the pasteurizer hold time, the cooling section would allow generation down to the

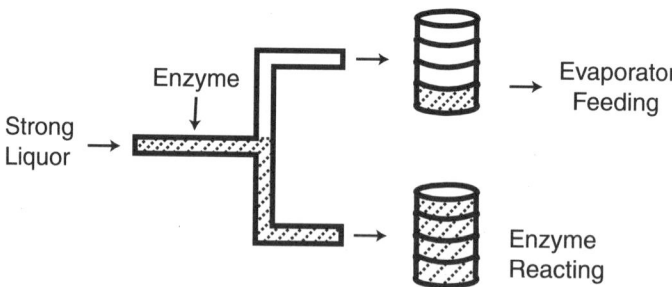

Figure 7.3 Flow scheme for addition of pectinase enzymes to pulp wash liquor for viscosity reduction, indicating evaporator feeding and enzyme reaction occurring.

temperature (30–40°C) of optimum activity of the pectinase used. The product is held at this temperature until the reaction is complete and then sent to the evaporator. Inactivation of the added pectinase and any of its microbial contaminants occurs in the evaporator.

Alternatively, one might use continuous metering of enzymes into a single tank feeding the evaporator. When the tank starts to fill, pectinase metering begins; when the tank is full, evaporator feed starts and sufficient time has elapsed to allow enzyme reaction. This method is based on the tank size and flow rate to accommodate the 0.5- to 1.0-hr average system reaction time in a continuous manner. The enzyme solutions may have proteolytic enzyme activity, which, if diluted or warmed to room temperature during the period of use, may result in self-digestion and loss of pectinase activity. Therefore, monitoring of the process must occur for best results.

Pectinases are used to facilitate juice and soluble solids recovery from the pulp of many fruits. It has been shown that treating citrus juice pulp with pectinases followed by countercurrent pulp washing may result in a 20% increase in recovery of soluble solids (Braddock and Kesterson, 1976, 1979). This was first suggested with a recommendation to allow the enzyme reaction to occur over a period of several hours (Villadsen and Möller, 1967); however, because of the opportunity of microbial fermentation and degradation in the raw pulp, hold times should not extend over 1 hr. Because a variety of substrates and preparation media are used to grow the organisms from which pectinases are isolated, the manufacturers should be consulted regarding the specific activities of the preparations. The citrus product manufacturer should also be aware that most of the enzyme preparations are from mold culture, are not sterile, and have the potential to contaminate the product with various types of microorganisms.

7.2.1.2 Pectinase Types Commercial pectinase users normally rely on the manufacturers for technical applications information. Considering specific applications, such as pulp wash viscosity reduction, individual enzyme preparations may contain a variety of specific enzymes, which can have desirable or undesirable effects on the product (Voragen et al., 1986). The mode of action of some pectinases is represented by the reactions shown in Figure 7.4. These enzymes reduce viscosity by decreasing the pectin molecule chain length, resulting in varying (shorter) chain length polymers of galacturonic acid residues. Pectinesterase (see Chapters 5 and 6) action results in cloud loss but is needed to produce the free carboxyl groups (pectic acid) necessary for activity of enzyme preparations high in endo- or exo-polygalacturonase. Citrus fruits have low natural amounts of polygalacturonases, with maximum activity in the flavedo abscisson zones near the stem > albedo > juice pulp (Riov, 1975). Problems with polygalacturonases occur as a result of the requirement for pectinesterase activity, which results in cloud loss and potential gel formation with the natural calcium ions in the pulp wash liquor. However, if the reaction time is optimized, polygalacturonase preparations may be successfully used for vis-

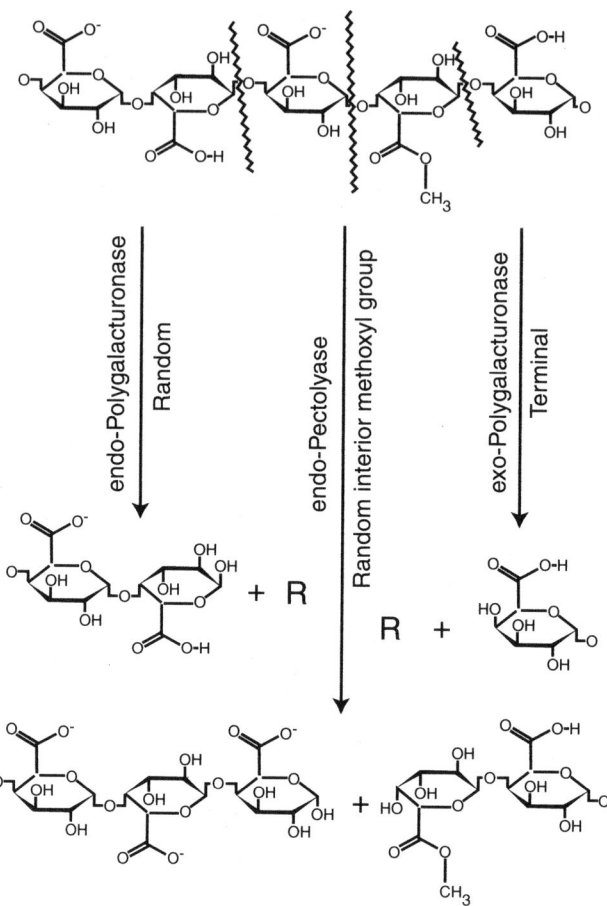

Figure 7.4 Mode of action, showing points of chain cleavage for different types of pectolytic enzymes.

cosity reduction. Some commercial preparations are high in pectin lyase activity, which reduces chain length by random cleaving next to interior galacturonic acid groups esterified with methanol (Fig. 7.4). Action of commercial preparations high in pectin lyase requires no prior pectinesterase in order to break the chains, resulting in more cloud and less gel problems in the product.

7.3 IN-LINE PULP WASH

Historically, in Florida, as in much of the world's citrus processing industry, blending pulp wash liquor with the juice feeding the evaporator has been legal. However, for a number of years during the 1970s–1994, this process was illegal in Florida. Now, although recent rules do not clearly state legality concerning this subject (Florida Department of Citrus, 1998), it has been established in Florida to allow so-called in-line pulp washing. The process consists of sending the strong liquor obtained from washing the pulp at the time of juice extraction to the juice stream en route to the evaporator. Process parameters are little different from the original descriptions (see previous discussion of FMC and Brown processes), with the result being a small dilution (near 10%) of the solids in the evaporator juice feed stream. It is important to state that a manufacturer will certify, if the buyer specifies, that juice concentrate does not contain pulp wash. Also, in Florida, continuous government inspection monitors products in the processing plants and provides certificates that a concentrate does not contain pulp wash. Additionally, in-line pulp wash is not permitted in NFC juices, and separately manufactured pulp wash concentrates must be in specifically identified containers with 50–100 ppm sodium benzoate added as a tracer (Florida Department of Citrus, 1998).

7.4 QUALITY ISSUES

7.4.1 Pulp Wash Concentrate

The major quality issues with pulp wash concentrate are related to the physical and chemical properties and the flavor. Extreme washing and extraction parameters, especially if high concentrations of rag, membrane, and core are components, may increase the water-soluble pectin, flavonoids, and insoluble solids content to the point of causing handling and concentration problems. This may be dealt with, in part, by use of pectinases. However, most manufacturers with experience in handling this product recognize the limitations of too rigorous washing and finishing. The same conditions also can extract undesirable bitter substances, above the level of good manufacturing practices for this product. Generally, product quality specifications are similar to those for concentrated orange juice. More exact product specifications are determined by agreement between users and the manufacturers. Pulp wash concentrates do not have the good flavor or color of true juice concentrates. Since most pulp wash color scores are poor, color measurements based on spectrophotometry have been studied in attempts to quantitate the amount of pulp wash added to true juice concentrates (Petrus and Dougherty, 1973).

Monitoring of temperature in the evaporator during concentration can be used as an indication of pulp wash quality, where high viscosity may be a problem. If the pulp is soft or fragile or excessive pectin is contained in the product, higher viscosities will be reflected by higher than normal evaporator process temperatures. Viscous liquids subjected to high temperatures may cause local overheating, poor heat transfer, and result in a cooked flavor and product browning. On the other hand, insufficient heating may not inactivate the added pectinases, and activity during product storage and handling can affect the cloudiness or gelation of the concentrate. Concentrate that had significant native pectinesterase activity prior to pasteurization (or pectinase addition) of the strong liquor can also gel or lose cloud during storage.

Typical quality parameters of pulp wash liquids and concentrates are described in the following statements. Pulp wash concentrate from oranges is usually between 60 and 65 °Brix, while grapefruits may be lower (50 °Brix). Viscosities to 10,000 mPa's (Brookfield viscometer, No. 4 spindle, 60 rpm, 26°C) and a No. 2 gel rating would be acceptable, above 15,000 mPa·s would be high. If the strong liquor was pectinase-treated prior to concentration, viscosities in the range of 2000–5000 mPa·s would be typical. Viscosities in this range will meet handling requirements for ease of mixing and blending by the finished product users. Astringent or bitter-flavored pulp wash concentrates may result from immature fruit or peel juices.

Pulp wash diluted to 10 °Brix from concentrate should have initial cloud of not more than 20% light transmission (%T) and commonly, 10%T. Cloud values after 24 hr at room temperature should be no greater than 35%T. Sinking pulp should be less than 10% and preferably 5% because particle precipitation is undesirable in beverages containing pulp wash. Separation of serum and cloud constituents over 4 hr in a graduated cylinder (100 mL) will normally result in 5–20 mL separation for pectinase-treated liquids. Serum viscosity of 10 °Brix liquid from a pectinase process is about 5 mPa·s with untreated liquid from 10 to 12 mPa·s (Braddock, 1980). Recoverable oil of 10 °Brix liquid reconstituted from concentrate is low, 0.005–0.015%. Dilute pulp wash normally contains no floating pulp and should be free of hesperidin crystals or other defects. Pectinesterase activity should be less than 0.5 PEU.

7.4.2 In-Line Pulp Wash

Due to considerable quantities of pulp wash used in concentrate manufacture in Florida after 1957, the federal standards of 1963 approved the process until the 1970s prohibition by Florida Department of Citrus regulations. This prohibition was a result of conditions related to quality improvement and economics, not because pulp wash solids were not considered juice solids. Since it was uneconomical for processors to send juice solids to the feed mill, separation of the in-line stream to make pulp wash concentrate became a reality. After commercialization of pulp wash concentrate, it was apparent that it was not quality equivalent to FCOJ, and there was the possibility it could be detrimental if mixed into this product in greater proportions than possible during in-line addition. When in-line pulp wash is added to concentrate, the issue dividing processors and growers is the economics of the total juice yield from the fruit and not quality of the product.

7.4.2.1 *Detection in FCOJ* Now that in-line addition of pulp wash is again legal in Florida, as has always been the practice in Brazil and other citrus processing locations, the issue of concentrate containing pulp wash has resurfaced as a matter of concern by some buyers of FCOJ in the world market.

The European Union (EU) requires orange juice to be made only by mechanical means from sweet orange (*Citrus sinensis*) (AIJN, 1996). In terms of either chemical detection or flavor, there should be negligible effects of in-line addition of pulp wash to FCOJ quality. The discussion of adulteration detection of this product is not new and was not particularly successful in the past using spectrophotometric methods (Petrus and Dougherty, 1973; Petrus and Attaway, 1980). A method for calcium concentration by a colorimetric procedure was established for pulp wash and extraction water (Coleman et al., 1985). Mineral composition and statistical neural networks have been used to compare orange juice and pulp wash concentrates, although results were inconclusive (Nikdel, 1991).

The new citrus variety, Ambersweet, may also make juice authenticity based on flavonoid analysis more difficult. The narirutin/hesperidin ratio of an orange juice containing Ambersweet (or navel) juice may appear to contain pulp wash (Widmer and Barros, 1995). Although Ambersweet is one-eighth grapefruit, no naringin was reported in the juice (Shaw et al., 1997). Those considering measurement of galacturonic acid from pectinase treatment to reduce pulp wash viscosity should understand that most in-line systems do not use (or need) pectinases, some enzyme preparations contain very little polygalacturonase, and citrus fruits do contain some of this enzyme. Viscosity is reduced by centrifuging. Generally, very few chemical changes would be expected, even if the juice itself were treated with pectinases (Braddock, 1981; Baker and Bruemmer, 1971). Although analytical techniques are today very sophisticated, the broad ranges of fruit and processing variables for both juice extraction and pulp wash recovery established above and in Table 7.1 are valid and should be considered when planning adulteration detection studies.

The chemist considering reliable detection of in-line pulp wash adulteration of FCOJ might consider the wide range of the above processing variables and the following arguments for detection of peel and pulp wash extracts, based on Australian orange juice studies (Johnson et al., 1995). These authors proposed use of the flavonoid-derived compound, phlorin, to detect peel extract or pulp wash addition to juice, after a continuous diffusion process for recovering soluble solids from peel and comminuted fruit (apples) was proposed for citrus (Gunasekaran et al., 1989). Phlorin is present in high concentration in the albedo > rag and membrane > juice and presumably, juice pulp (Louche et al., 1998).

Johnson et al. (1995) used the example of detecting a "worst-case," 5% addition of 10 °Brix peel extract (200 mg/L phlorin) to juice (3 mg/L phlorin). These values represent low amounts of peel extract phlorin and high amounts in unadulterated juice. At 5% dilution with peel extract, the detected phlorin content was three times greater than the amount found in unadulterated juice.

The phlorin concentration of pulp wash samples calculated at 10 °Brix was 110 mg/L (Johnson et al., 1995). Using the same dilution ratio (5%) as above, the phlorin content of juice containing 10 °Brix pulp wash would be only about 1.5 times the amount in juice alone. If one applies this reasoning to a discussion of detecting addition of in-line pulp wash to juice, it would be difficult, if not impossible, because in-line pulp wash strong liquor may only be 5 °Brix, containing 0.5 times the phlorin content at 10 °Brix. Even detecting the addition of peel extract at 5 °Brix would be questionable. A practical approach verifying this argument lies in the dilution limitation imposed by considering mass balance data for Valencia orange juice and in-line pulp wash recovery. Valencia orange juice recovery yields 6.5 lb soluble solids/box (Table 2.2), while pulp wash solids recovery is 0.5 lb soluble solids/box (Table 7.1). Based only on the soluble solids content, this dilution is about 8% (0.5/6.5), close to the 5% suggested by Johnson et al. (1995).

The above arguments have the following basis in calculation:

Given: 5% addition of 5 °Brix pulp wash containing 50 mg/L phlorin
          In-line pulp wash is 5 °Brix
          Juice contains 2–4 mg/L phlorin (amount in the juice pulp).
Calculate: Phlorin resulting from 5% pulp wash addition to juice

Maximum in juice: $0.950 \text{ L} \times 4 \text{ mg/L} + 0.050 \text{ L} \times 50 \text{ mg/L}$

$$= 6.3 \text{ mg phlorin}$$

Minimum in juice: $0.950 \text{ L} \times 2 \text{ mg/L} + 0.050 \text{ L} \times 50 \text{ mg/L}$

$$= 4.4 \text{ mg phlorin}$$

Soluble solids balance:

Valencia orange juice = 6.5 lb soluble solids/box (seasonal average)
Pulp wash soluble solids = 0.5 lb soluble solids/box (Table 7.1)
Maximum yield in-line pulp wash added to juice = 0.5/6.5 = 7.7%
(In-line pulp washing may only recover 80% of the 0.5 lb soluble solids =
6.2%)

An actual plant scenario (FMC plant) might be as follows:

Juice: 16 extractors × 2 boxes/min/extractor × 6.5 lb soluble solids/box =
208 lb soluble solids/min
Pulp wash: 16 extractors × 2 boxes/min/extractor × 0.5 lb soluble solids/
box × 0.80 = 12.8 lb soluble solids/min

Thus, as above, 12.8/208 = 6.2% from pulp wash added to juice.

Considering the previous discussion and all the processing and fruit variety variables, detecting in-line pulp wash would be difficult. Also, even if a data comparison was statistically performed at the 2-mg/L low phlorin value in the juice, unless an actual sample of the juice without pulp wash was collected, there would be no way to determine the actual phlorin level. There are also commercial juices with very high pulp content, some with pulp levels of 25%. Perhaps the only hope for pulp wash adulteration detection is for chemists to discover a chemical present in the peel, rag, membrane, and pulp not present in the juice. Another suggestion is to drop objections to use of in-line pulp wash.

7.4.2.2 Water Quality The quality of water used for washing pulp has been an issue of concern for some fearing that use of well water adds something not naturally from the fruit. However, every consumer who dilutes FCOJ with tap water adds the chemicals present in the locale of the water to the juice. Use of juice evaporator condensate water for pulp washing has seen limited practice, primarily because some European countries may have felt it was more natural than potable well water. In reality, condensate water used for fruit cleaning and pulp washing is probably not a good idea since it may be readily contaminated by the environment from its origins in the evaporator to the storage tanks where it is accumulated. A good study has identified some very thermophilic bacteria, which may commonly contaminate fruit, soil, and evaporator condensate water used for pulp washing (Wisse and Parish, 1998). The spore-forming species of *Alicyclobacillus* may grow in the acid environment of citrus juices and produce bad off-flavors in the packaged product held in storage, particularly in Europe where ambient storage temperatures are used. Condensate waters from a wide range of processing plants at different geographical locations were shown to be contaminated with these organisms.

7.5 UTILIZATION

Manufacturers who wish to add fruit solids and natural cloudiness to juice, beverages, and drink products use pulp wash concentrates. Usually cheaper than juice concentrate, this product imparts more natural cloudiness to beverages and is usually listed in the label statement as concentrated orange juice or solids. An advantage of this use is that synthetic ester gum and resin clouds are not necessary, which has marketing appeal to consumers leery of label declarations. A disadvantage of using pulp wash concentrate in beverages is the precipitation of coagulated, fine pulp particles as the product is allowed to stand undisturbed for long periods (Sutherland, 1979). Another defect of pulp wash concentrate used in beverage bases is the presence of hesperidin flakes. This material builds up as scale on the tubes in the evaporator and then dislodges as small pale yellow flat pieces, which may be observed after dilution of the concentrate with 10 parts water. More than three to four flakes on the bottom of a beaker would be considered excessive and are undesirable defects in finished beverages.

7.6   PULP RECOVERY

7.6.1   Process

During juice recovery, the extracted juice passes immediately to finishers for removal of part of the ruptured juice vesicles, or pulp. This pulp may be recovered as a by-product from the pulpy juice streams from either Brown or FMC extractors. The object of the recovery process is to manufacture large pulp sacs from the pulpy juice stream, with most of the juice removed. This product preferably contains particles capable of being retained on a 20-mesh screen with a high percentage of floating pulp. Pulp recovery from both processes is similar, once the pulpy juice stream is isolated from the extractors or finishers. The amount of pulp recovered can be near the amount stated in Table 7.1 but because of quality concerns may be only half as much. Since the unwashed product is preferred, pulp recovery lowers juice and pulp wash solids yield as discussed above.

A flow diagram for the process is presented in Figure 7.5, where part of the pulpy juice from the juice finisher first goes to a screw or paddle finisher for classification or removal of some defects such as seeds and pieces of rag. The pulpy juice then passes to a conical cyclone separator, called a hydroclone, which further classifies the pulpy liquid stream by separating the embryonic seeds and small defects. These discharge out the bottom of the hydroclone and the pulpy juice stream (5–6 parts juice:1 part pulp) passes through the top to a paddle finisher (or a screen) for concentration of the pulp to >500 g/L. The hydroclone operates by pumping the stream at high enough pressure to allow sufficient spiral, tangential flow of the pulpy juice along the walls and then up and out through the center. The heavier embryonic seeds and defects collect on the walls, flowing downward and are discharged through a valve opening in the bottom. The next step involves pasteurization at near 90°C for 0.5–1.0 min for microbe and enzyme inactivation, followed by chilling to 2–5°C. Because of difficulties encountered pumping the concentrated pulp, heat exchangers with large horizontal tubes and spiral baffles must be used for efficient pasteurization (Morris, 1996). A final finisher recovers excess juice and dries the pulp before product packaging into 5-gal cartons, 55-gal drums, or bag-in-box containers. Cartons and drums are usually stored frozen, while bags may be kept refrigerated for a short period, if not packaged aseptically (Morris, 1996).

7.6.1.1 FMC Pulp   The FMC system has two types of pulp recovery (standard or premium) determined by the juice extractor components, with pulp difference primarily related to particle size. Standard pulp from standard juice extractor components results in smaller size pulp particles from the 0.040-inch perforations of the extractor strainer tubes. Premium pulp recovery most importantly uses extractor strainer tubes with vertical slots in place of perforations to yield particles of large uniform size, free from defects, and having a high

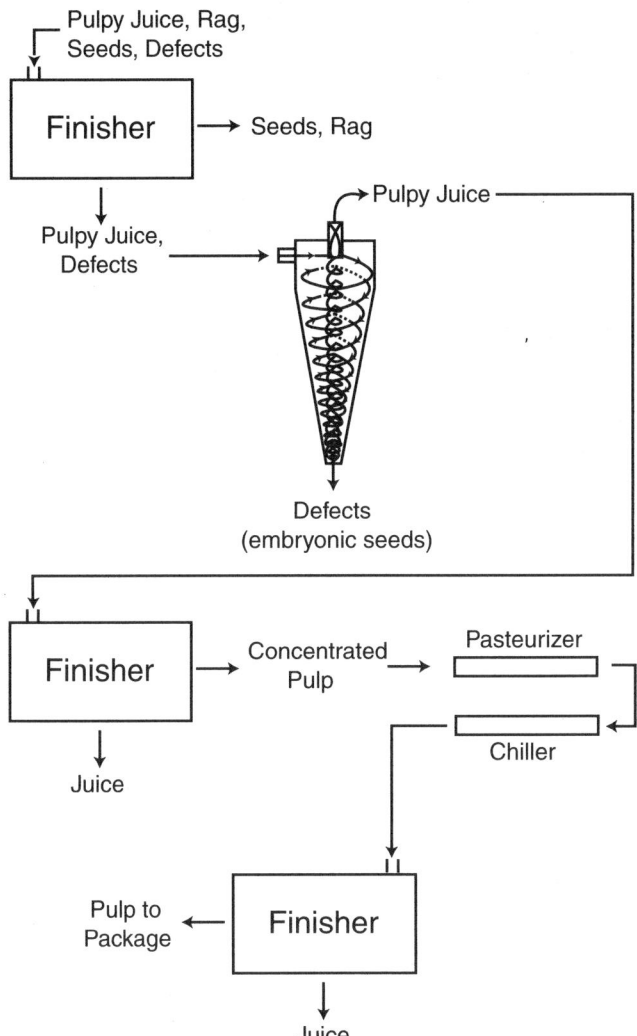

Figure 7.5 Flow diagram of a pulp recovery process illustrating isolation of pulp, defect removal by hydroclone, pasteurization, and packaging.

percentage of floating vesicles (Johnson, 1987). This process treats the pulp more gently and results in a higher percentage of large floating sacs than the standard process.

7.6.1.2 *Brown Pulp* Part of the pulpy juice stream from juice finishers in the Brown juice extractors may also be used for pulp recovery. The juice reaming process favors production of large floating vesicles in the pulpy juice stream. The pulpy juice stream is sent directly to hydroclones for removal of defects and embryonic seeds, concentrated by finishing to reduce the volume to be pasteurized, heat treated, and chilled, and the final pulp is packaged (Waters, 1985).

7.6.2 Utilization

The primary use for pulp is blending with concentrates and bases to add texture, body, and pulpy character to reconstituted juices and drinks. An amount of pulp floating at the top of juice poured into a glass makes the impression the product is "natural," freshly squeezed juice (Sutherland, 1979). The product is also used in juice-containing beverages and high-pulp juices, such as home-style juices, that have marketing appeal to consumers. Quality measures determine the weight of pulp remaining on a 20-mesh screen from pouring 24 oz. reconstituted juice (6 oz. concentrate). This is commonly 4–8 g pulp/6 oz. can, but may be as high as 25–30 g/L for reconstituted high-pulp products that imitate natural products. Addition of pulp to product can result in problems controlling defects such as increased sedimentation of embryonic seeds, seed fragments, hull pieces, rag, core, and even sand. Microbiological problems can also occur if the product has been abused during manufacture or thawing, as yeast and molds can readily utilize the sugars in the product. Thus, the pulp needs to be carefully monitored before use and during manufacture (Kolb, 1993). Quality evaluation is not very sophisticated, depending largely on visual methods, sometimes with magnification. One looks for large, uniform, unbroken pieces with few defects.

Frozen unwashed pulp containing natural juice solids is the preferred material for product manufacturing uses. A chopper capable of handling frozen blocks or drums is useful for rapid blending to minimize fermentation during thawing of containers. The energy and time to freeze 55-gal drums of pulp has not been published but should be near that for citrus sections, packed similarly. It was determined that ambient temperature drums of grapefruit sections in a $-22°C$ cold room took approximately 120 hr to freeze to within $\pm2°$ of the freezer temperature (Chen et al., 1984). Small commercial quantities of frozen pulp have been produced cryogenically by evenly spreading pulp over a moving belt and spraying with liquid (or dry) CO_2. The rapidly frozen product has individual pieces of juice sacs, which can be packaged and kept frozen for product applications (Kesterson and Braddock, 1976). When frozen product is thawed and added to product, because of the dryness of the material, it will absorb juice and swell in the reconstituted juice. If the pulp is weighed prior to addition, then screened from the juice and weighed, the resultant product will increase in weight due to the absorbed juice. This is a matter dealt with in the quality control laboratory of the processor or blender, to determine the proper addition values in the final product. A helpful table describing the amount of pulp in the finished product related to the weight and °Brix of the concentrate and added pulp has been published (Waters, 1985).

Citrus juice pulp has been proposed as constituents of several food products for which patents have been issued. Certain edible, cylindrical-shaped foods and a base useful for preparation of jams, gel-like desserts, or glaze coatings have been described (Blake, 1980a,b). Homogenized pulp also has been proposed to replace emulsified lipids for stabilization of cake frostings, preventing air and liquid loss from the frosting foam (Blake, 1980c). Frozen pulp has also been used as a component of aerated, soft-serve frozen fruit desserts (Blake, 1981). These and other applications illustrate that many potential uses exist to incorporate the by-product, pulp, as a functional ingredient in a variety of foods (Braddock, 1983).

REFERENCES

AIJN (1996). Code of Practice for Evaluation of Fruit and Vegetable Juices, AIJN, Brussels, Belgium.

Baker, R. A. and Bruemmer, J. H. (1971). Enzymic treatment of orange juice to increase cloud and yield, and decrease sinking pulp levels, *Proc Fla. State Hort. Soc* **84**:197–200.

Ballentine, P. L. and Ferguson, R. (1985). "Water extracted soluble orange solids: pulp wash." In Proc. 25th Ann. Short Course for the Food Industry, R. F. Matthews, Ed., FSHN Dept., Univ. FL, Gainesville, FL, pp. 233–247.

Beasley, L. (1998). Private communication, FMC FoodTech, Citrus Systems Division, Lakeland, FL.

Belk, W. C. (1964). Pulp washing with FMC inline extractors, *Trans. Citrus Eng. Conf.* **10**:19–36.

Blake, J. R. (1980a). Food product containing orange citrus juice vesicle solids, U.S. Pat. 4,205,093.

Blake, J. R. (1980b). Cooked comestible base containing citrus juice vesicles, U.S. Pat. 4,232,053.

Blake, J. R. (1980c). Citrus juice vesicle containing frosting compositions, U.S. Pat. 4,232,049.

Blake, J. R. (1981). Non-dairy, aerated frozen dessert containing citrus juice vesicles, U.S. Pat. 4,244,981.

Braddock, R. J. (1980). "Quality of citrus specialty products: Dried pulp, peel oils, pulp wash solids, dried juice sacs." In *Citrus Nutrition and Quality*, S. Nagy and J. A. Attaway, Eds., ACS Symposium Series No. 143, Washington, D.C., Chap. 12, pp. 273–288.

Braddock, R. J. (1981). Pectinase treatment of raw orange juice and subsequent quality changes in 60 °Brix concentrate, *Proc. Fla. State Hort. Soc.* **94**:270–273.

Braddock, R. J. (1983). Utilization of citrus juice vesicle and peel fiber, *Food Technol.* **37**(12):85–87.

Braddock, R. J. and Kesterson, J. W. (1976). Enzyme use to reduce viscosity and increase recovery of soluble solids from citrus pulp-washing operations, *J Food Sci.* **41**:82–85.

Braddock, R. J. and Kesterson, J. W. (1979). Use of enzymes in citrus processing, *Food Technol.* **33**(11):78, 80, 81, 83.

Chen, C. S., Ting, S. V., and Hill, E. C. (1984). Predicting temperature changes in freezing of citrus sections contained in steel drums, *Trans. Am. Soc. Agr. Eng.* **27**(5): 1604–1609.

Coleman, R. L., Saunders, M. S., and Baker, R. A. (1985). Calcium determination in citrus pulp wash—a comparison of a colorimetric procedure with atomic absorption spectrophotometry, *Proc. Fla. State Hort. Soc.* **98**:218–219.

Florida Department of Citrus (1998). Water extracted soluble fruit solids, Official Rules Affecting the Florida Citrus Industry. Ch. 20-64.021, March 19, pp. 17–19.

Gunasekaran, S., Fisher, R. J., and Casimir, D. J. (1989). Predicting soluble solids extraction from fruits in a reversing, single screw counter current diffusion extractor, *J. Food Sci.* **54**(5):1261–1265.

Johnson, T. M. (1987). Citrus pulp recovery, *Citrus Ind. Mag.* **68**(8;9):36–39, 17–18.

Johnson, R. L., Htoon, A. K., and Shaw, K. J. (1995). Detection of orange peel extract in orange juice, *Food Australia* **47**(9):426–432.

Kesterson, J. W. and Braddock, R. J. (1976). By-products and specialty products of Florida citrus, Fla. Agr. Exp. Sta. Tech. Bull. No. 784. University of Florida, Gainesville, FL, p. 87.

Kimball, D. A. (1991). "Citrus juice pulp." In *Citrus Processing Quality Control and Technology*, Van Nostrand Reinhold, New York, Chap. 7, pp. 102–116.

Kilburn, R. W. and Dillman, C. A. (1951). A method for estimating the amount of citrus juice in finisher pulp, *Proc. Fla. State Hort. Soc.* **64**:138–140.

Kolb, E. (1993). Citrus juice finisher pulp processing and its use in fruit juices, *Flüssiges Obst.* **60**(1):6–8.

Louche, L. M., Gaydou, E. M., and Lesage, J. C. (1998). Determination of phlorin as peel marker in orange (*Citrus sinensis*) fruits and juices, *J. Agric. Food Chem.* **46**(10):4193–4197.

McKinnis, R. B., Andrews, R. A., and Jones, H. L. (1964). Pulp wash system development, *Trans. Citrus Eng. Conf.* **10**:1–18.

Morris, C. E. (1996). New technologies add value to pulp and by-products, *Food Eng.* **68**(9):157–158, 162, 164, 166.

Nikdel, S. (1991). Orange juice and pulp wash concentrate analyses with ICP-AES and their characterization via neural network technology, 42nd Ann. Citrus Processors' Mtg. Abstracts, Oct. 17, University of Florida, CREC, Lake Alfred, FL, pp. 23–26.

Olsen, R. W., Wenzel, F. W., and Huggart, R. L. (1958). Recovery of fruit solids from orange pulp, *Proc. Fla. State Hort. Soc.* **71**:266–274.

Petrus, D. R. and Attaway, J. A. (1980). Visible and ultraviolet absorption and fluorescence excitation and emission characteristics of Florida orange juice and orange pulpwash: detection of adulteration, *J. Assoc. Off. Anal. Chem.* **63**(6):1317–1331.

Petrus, D. R. and Dougherty, M. H. (1973). Spectrophotometric analyses of orange juices and corresponding pulp washes, *J. Food Sci.* **38**:913–914.

Riov, J. (1975). Polygalacturonase activity in citrus fruit, *J. Food Sci.* **40**:201–202.

Shaw, P. E., Moshonas, M. G., and Hearn, C. J. (1997). "Chemical analysis in classifying Ambersweet citrus hybrid as an orange for juice processing." In *Chemistry of*

Novel Foods, A. M. Spanier, M. Tamura, H. Okai, and O. Mills, Allured Publishing Corp., Carol Stream, IL, Chap. 11, pp. 135–152.

Sutherland, R. (1979). "Utilization of specialty citrus products." In Proc. 19th Ann. Short Course for the Food Industry, R. F. Matthews, Ed., FSHN Dept., Univ. FL, Gainesville, FL, pp. 143–156.

Villadsen, K. J. S. and Möller, K. J. (1967). Processing citrus fruits, U.S. Pat. 3,347,678.

Voragen, A. G. J., Wolters, H., Verdonschot-Kroef, T., Rombouts, F. M., and Pilnik, W. (1986). Effect of juice-releasing enzymes on juice quality, *Int. Fruchtsaft-Union Symposium. Den Haag.*, **19**:453–462.

Waters, R. (1985). "Pulp recovery and packaging system." In Proc. 25th Ann. Short Course for the Food Industry, R. F. Matthews, Ed., FSHN Dept., Univ. FL, Gainesville, FL, pp. 209–232.

Webb, E. L. (1997). Describing a citrus pulp wash system using a modified McCabe-Thiele method, M.S. Thesis, Dept. Chem. Eng., Univ. So. FL, Tampa, FL.

Wenzel, F. W., Huggart, R. L., and Olsen, R. W. (1959). Water extraction of fruit solids from orange pulp, *Trans. Citrus Eng. Conf.* **5**:1–8.

Widmer, W. W. and Barros, S. (1995). Flavonoids in Ambersweet orange and the impact on juice adulteration detection, 46th Annual Citrus Processors' Meeting Abstracts, Oct. 11, University of Florida, CREC, Lake Alfred, FL, pp. 21–26.

Wisse, C. A. and Parish, M. E. (1998). Isolation and enumeration of sporeforming, thermoacidophilic, rod-shaped bacteria from citrus processing environments, *Dairy, Food and Env. Sanitat.* **18**(8):504–509.

CHAPTER 8

WHOLE JUICE CELLS, DRUM-DRIED JUICE SACS, AND SECTIONS

The products from the edible portion of the fruit have obvious functional uses in foods. As discussed in the last chapter, finisher pulp and pulp wash are recovered because they have economic value and functionality in juices, juice-containing drinks, and beverages. Interest in the pulp stems from its location in the edible tissue, making it free of bitterness and impurities found in the peel. The pulp has a mild citrus flavor and texture and with certain processing techniques, can be used in other foods.

8.1 WHOLE JUICE CELLS

A commercial process was described in the 1940s to manufacture whole juice vesicles, Juicells, from grapefruits (Anon, 1947). The process, by today's standards, was labor intensive, consisting of loosening the peel by soaking fruit in hot water, coring, halving, removing the juice vesicles on a rotating wire-loop wheel, inspecting for defects, pasteurizing, and canning in sweetened juice. Uses for the product included a breakfast dish, fruit cocktail base, and as a dessert.

The author studied a similar process to recover whole juice vesicles from grapefruits and lemons with the objective of minimizing the labor required (Braddock, 1983). The approach required using commercial citrus peeling and sectionizing equipment as a first step, followed by separation of the juice-containing vesicles. One technique involved loading the sections into a front-loading clothes washing machine, with the agitation and rinsing cycles breaking the pulp loose from the membrane and core. Results also indicated that heating the segments to 80°C (176°F) in the presence of 1–2% citric (or phosphoric) acid helped melt the lipid cementing the juice vesicles together (Shomer et al., 1975) and dissolve the carpellary membrane. This allowed the juice vesicles to float free in the mixture, where they were recovered by passing over a vibrating screen. A theoretical yield of 65% of the fruit mass was determined by hand-peeling and careful separation of whole juice cells from grapefruits.

There are commercial products such as yogurt and drinks, containing whole juice cells manufactured by processes, that generally require considerable hand labor to scrape the vesicles from the carpellary membrane. Production losses occurring during peeling, segment separation, pasteurization, and filling result in final yields of 40–45% because when cells break, the juice content leaks out. Thus, a number of devices have been patented seeking to mechanize the process. One of these uses fluid (or air) jets to break the vesicles loose from cut half-portions of tangerines, separating the product with a screen (Ifuku et al., 1981). A process to increase quality and recovery efficiency of whole juice sacs by freezing sections between −35 (−31°F) and −120°C (−184°F) and breaking the frozen samples into individual sacs by impact has been published (Watanabe et al., 1987). Another apparatus separates juice sacs from cryogenically frozen fruit by crushing the fruit and recovering the juice sacs from the frozen crushed fruit (Ando et al., 1988). Other processes seek to increase mechanization of juice sac recovery through use of a rotating screen drum containing a counter-rotating coaxial shaft with fingers to separate juice sacs from sections (Kolodesh, 1989a) or by coring fruit held in cups and recovering the sections with the juice vesicles (Kolodesh, 1989b). There have also been other patents for devices to aid in recovering juice-containing vesicles by impinging high-pressure water on cut fruit (Kock et al., 1990, 1991). Once separated into clusters of juice sacs, slurries are pumped in streams until they segregate into individual vesicles (Watanabe, 1985). The main disadvantage of these processes is that they still involve the labor-intensive process of peeling fruit and producing the sections, prior to any recovery of the juice vesicles. Mechanical sectionizers have the problem of sections breaking in the vibrating screen de-seeders, allowing juice cells to fall through, which reduces yield. Research is needed to establish efficient production of whole, unbroken cells and optimize processing, packaging, and storage quality for this product. An interesting possibility, which could revolutionize production of this product, has been proposed. Juice vesicles have been grown by tissue culture in sterile media from grapefruits, lemons, and oranges (Tisserat et al., 1989). Such an approach might lead to a viable commercial method to reduce, in part, the high labor cost of this interesting by-product.

8.2 DRUM-DRIED JUICE SACS

As discussed in Chapter 7, juice sacs or pulp can amount to 10–20% of the total fruit residue after juice recovery. Besides recovery and use as frozen pulp, separation and dehydration have been used to offer utilization in some foods requiring a dried raw material.

8.2.1 Processing

Although dried juice sacs may be manufactured from either washed or unwashed juice sacs, the presence of invert sugars results in a dark brown color and a caramel flavor in the unwashed dried product. Washing through a conventional pulp washing process removes most soluble sugars, allowing effective drying. The concentrated soluble sugars are sold as pulp wash or blended inline to make concentrated juice. Raw material pulp yields for oranges and grapefruits were presented in Table 7.1.

Freeze-drying and foam-mat drying have been used to prepare dried unwashed juice sacs; however, the cost is very high, and there has been little demand for the product. Citrus peel dryers have also been used to dry washed juice sacs, although the high moisture content of the pulp results in sticking and burning on hot surfaces in the dryer. This yields black pieces in the product, even when product temperatures are below 60°C. Most commercial dried juice sacs have been produced from washed juice sacs by steam drum dryers. During this process, washed pulp is applied to rotating stainless steel drums, heated from the inside by steam. The raw pulp is usually not pasteurized prior to application to the drums because the drum temperature and low water activity of the finished product inhibit microorangisms and enzyme activity. Drum speed, temperature, and pulp layer thickness is individually regulated to yield product below 10% moisture content.

Dried juice sacs have been commercially manufactured several times since the 1950s in the citrus industry and the process is not complicated. The product was reinvented in the 1970s when consumer interest in dietary fiber sources became popular. A dietary fiber product containing juice pulp, membranes, and peel was prepared by first mixing with sesame flour, grinding, and dehydration to 7% moisture (Lynn, 1980). Another juice sac fiber product was produced from pulp by pH adjustment above 4.0, then freeze drying to facilitate rehydration and improve the textural properties of the dried material (Eisenhardt et al., 1984).

8.2.1.1 Yield Although it is difficult to produce a good quality product, the yield of unwashed dried juice sacs can vary from 0.3 to 1.4 lb/box of fruit, depending on finishing and recovery. Yield of dried and washed is one-third of unwashed juice sacs and may range from 0.1 to 0.5 lb/box, with an average yield of approximately 0.4 lb/box (Kesterson and Braddock, 1973). A typical material balance for processing and drying orange and grapefruit juice sacs is presented in Table 8.1. These values are from the author's research (Kesterson and Braddock, 1973) and are comparable to later studies (Ferguson and Fox, 1978; Passy and Mannheim, 1983). Generally, after washing to remove the sugars, the washed pulp is near 95% moisture, and the slippery texture makes

TABLE 8.1   Mass Data for Processing Quantities of Dried Orange and Grapefruit Juice Sacs from Unwashed and Washed Raw Juice Sac Material from 10 Boxes of Fruit

Process Stream	Orange	Grapefruit
Unwashed sacs (kg)	19.6	16.7
Dry unwashed sacs (kg)[a]	2.1	1.7
Washed sacs (kg)	14.6	14.6
Dry washed sacs (kg)[a]	1	1

[a]Dry sacs are at 10% moisture.

mechanical dewatering by pressing or finishing below 90% difficult. A complete engineering feasibility study of dried juice sac manufacture indicated that finishing could lower the moisture content of the raw washed pulp to near 90%, such that a drying ratio of 9/1 could be achieved at 10% moisture product (Ferguson and Fox, 1978). More efficient washing to remove soluble solids results in higher drying ratios; while more solids in the raw material increased yield because less water was removed to attain the product moisture content (Passy and Mannheim, 1983).

8.2.1.2   Properties   The dried product is scraped from the drum with a steel blade as thin sheets of dried juice sacs, which can be milled to any size required for the intended application. The form may be flakes, grits, or flour. Flakes have the appearance of oatmeal and are prepared from the drum dryer sheets by grinding in a Hobart mill to pass a 4- to 8-mm sieve opening. A hammer mill is used to grind dried flakes into grits, similar to corn meal, having a particle size range between 0.5 and 1.0 mm, and flour is milled to particles of 0.15–0.2 mm diameter.

Dried juice sacs from either oranges or grapefruits have a mild, bland flavor and aroma that is not characteristic of the fruit source. The color of product from oranges is light orange when fresh but soon fades when exposed to light and air. Product from grapefruits is white, unless red grapefruits are used; then the pink color bleaches rapidly with storage time. Even storage in opaque, air-free containers will not preserve the color, and after a number of weeks, it is impossible to distinguish either product. Like many dried foods, the large surface area available for reaction with O_2 results in difficulty of maintaining initial texture and flavor quality. Auto-oxidation of the cell-wall lipids and pigments is considered the reason for dried product degradation resulting in a haylike flavor during storage. Some shelf life extension may be expected by antioxidant treatment prior to drying, careful packaging to minimize air exposure, and storage at cool temperatures. An antioxidant with good carry-through (e.g., butylated hydroxyanisole) for the drying process is necessary.

8.2.1.3 Composition   The chemical and physical compositions of dried juice sacs are listed in Table 8.2. These values are useful for considering blending with foods from nutritional and functional purposes. The composition of the alcohol-insoluble solids (AIS) of juice vesicles of oranges, grapefruits, and tangerines including monosaccharides, hemicellulose, cellulose, pectin, fat, protein, and ash has been published (Ting, 1970). Interest in dietary fiber also resulted in publication of other reports characterizing the fiber and protein of citrus juice vesicles on a fresh weight basis (Braddock and Graumlich, 1981). Also, fiber, vitamins, and other properties were studied from washed, drum-dried pulp (Porzio and Blake, 1983), as a percentage of the AIS (Ting and Rouseff, 1983) and of juice pulp refuse from mandarins (Yoshida and Ueda, 1984). Of these studies, total pectin content and crude fiber were the most significant components in these products on a moisture-free basis. Thus, dried juice sacs should be considered as a useful source of soluble and insoluble dietary fiber.

8.2.1.4 Handling   Dried juice sacs have been packaged in lined fiber drums or cartons, which may be warehoused at ambient temperature. Sanitary Good Manufacturing Practice (GMP) manufacturing, packaging, and storage environments must be adhered to because of the intended food applications. Material with moisture contents above 15% may ferment or mold during storage, and even correctly dried product may contaminate food products to which it is added. The heat of drum drying can be expected to reduce microorganism numbers, but contamination may occur during packaging. Insects can also contaminate the dried product during storage, as both weevils and moths have caused product losses.

TABLE 8.2 **Chemical and Physical Properties of Washed Dried Juice Sacs from Oranges and Grapefruit**

Property	Orange	Grapefruit
Moisture (%)	8–10	8–10
Crude fiber (%)	18	16
Protein (%)	7–9	7–9
Lipid (%)	2–3	1–2
Ash (%)	3	2–3
Pectin (%)	21	26
Density (kg/m^3)	150	144
Water binding[a]	10/1	14/1
Fat binding[a]	4/1	4/1

[a]Ratio is the weight of water or fat absorbed to amount of dry material.

8.2.1.5 Utilization Dried juice sacs have been used in a number of food products and proposed food processing applications. The most common application appears to be as a functional food ingredient for the purpose of imparting textural properties associated with water binding and retention. With proper processing and modification, the dried juice sacs hydrate readily and have other benefits, such as an ability to mute flavors in some food products (Porzio and Blake, 1983). The product has been described as being cream-colored, very lightweight, hygroscopic, and flavor free, with applications and formulas proposed as an ingredient to make improved texture cookies and meatloaf (Hannigan, 1982). A list of potential functional ingredient uses for dried juice sacs for a number of foods was proposed as follows: provide pulp for dry beverage mixes; as a thickener or gelling agent for pie fillings, jams, sauces, and gravy; as bulk fiber source in breads, cake, cookies, and cereals; as a binder, texturizer in comminuted meats such as bologna; and as a filler for low-calorie foods, such as salad dressing (Ferguson and Fox, 1978). Dried juice sacs were reported to improve texture, height, and moisture retention of baked brownies (Feeney et al., 1988). Dried juice sacs milled to a particle size preferably between 30 and 50 μm have also been proposed as an ingredient to increase the dietary fiber content of fruit juices with minimal changes in juice viscosity, off-flavor, or gritty taste (Mills and Tarr, 1992).

While there are many proposed uses for dried juice sacs, there are some limitations to large-scale commercial utilization. It should be noted when considering baking applications that require blending dried juice sacs with flour from grain sources that the major component is pectin. If the functional use requires pectin, it would facilitate formulation to simply add pectin. There has been at least one report that contradicted some of the above-mentioned usefulness of adding juice sacs to baked foods. Careful study of processing and mixing conditions indicated that rapid hydration in products mixed with small amounts of dried juice sacs caused dough or batters to set too quickly (Blake, 1980). This resulted in nonelastic masses, which were unsuitable for bread or cake manufacture, unless the pectinaceous material was modified, or coated, with lipid.

Another limitation to commercial utilization of dried juice sacs has been the operational and capital cost of the drying process. After washing and finishing, the raw material at 90% must be dried to at least 10% moisture to make a stable product. This cost is not economical for food product applications, which have competing cheaper ingredients with similar functional properties. However, there may still be future applications for this product.

8.3 CITRUS SECTIONS AND SALADS

Processing traditional orange and grapefruit sections has been described previously (Chace et al., 1940; USDA, 1962; Kesterson and Braddock, 1976). Processing and product quality standards have also been outlined more recently

(Kimball, 1991). The cost of the process is high because of the hand labor required to produce a satisfactory product, so there are only a few remaining operations in the United States. However, at low cost, there is good demand for both orange and grapefruit sections as well as product containing other fruits, for example, pineapples, melons, and grapes.

8.3.1 Manufacture

Very little modernization of the process for recovery of sections has occurred since the early published descriptions (Chace et al., 1940; USDA, 1962). The first requirement is for properly mature, clean fruit, which may be hand-peeled, lye-peeled, or perhaps mechanically peeled.

8.3.1.1 Peeling Traditional hand peeling simply involves slicing the peel and albedo from the fruit with a knife, resulting in yield loss, but higher quality (Kimball, 1991). The most common peeling method involves steam or hot water (100°C) heating of the fruit for 5 min to loosen the peel for easy hand removal. Small amounts of adhering albedo may be removed by hand, but most is removed by lye treatment. A hot (88–100°C) lye solution (0.5–2.5% NaOH/ Na_2CO_3) is sprayed on the peeled fruit for about 12 sec, allowed to react to dissolve albedo and the outer membrane for 20 sec and rinsed with water for 30–45 sec (USDA, 1962). The rinsed, peeled fruit is conveyed to the sectionizing line.

Mechanical citrus peeling machines have long been a dream of inventors. An early rotating drum, scarifying machine tested in Florida could operate with one person and remove peel from 2000 kg of oranges or 3000 kg of grapefruits/ hr (Hood and Russell, 1916). A commercial peeler was manufactured by FMC (FMC FoodTech Citrus Systems Division, Lakeland, FL) in the 1960s. This machine had some limited success for peeling grapefruits. Another machine, used on a limited commercial basis, could be set up to peel oranges or grapefruits (Webb and Houghton, 1976). These machines, like others designed to peel citrus (Vincent et al., 1972), involved hand labor and for this reason, were slow and less economical than steam-hand-lye processes.

Another approach to removing the peel has been to take advantage of vacuum or pressure infusion of water, enzymes, or other fluids to soften the tissue, allowing removal. This idea was first commercialized around 1900 for syrup infusion preparation of candied peel from lemons and oranges (Elsbury, 1932). Vacuum infusion of enzymes to soften peel before hand removal has been described and commercialized on a limited scale (Bruemmer et al., 1978; Bruemmer, 1981). This process results in softening the peel by vacuum infusion of water containing a pectinase solution, whereby the pectinase helps loosen the albedo adhering to the outer segment wall (Adams and Kirk, 1991). When the peel is pulled off, the fruit surface is relatively free of albedo and visually attractive (Baker and Wicker, 1996). A very similar vacuum process claims that the enzyme treatment is not necessary, loosening the peel simply by vacuum or pressure infusion treatment of fruit with water (Pao et al., 1996; 1998a). A process combining steam treatment followed by a rapid decompression–vacuum treatment and mechanical roller removal of the peels has also been proposed (LeGrand and Stevens, 1991).

The infusion processes have a number of flaws. First, hand labor is still required to remove the peel. Fruit maturity and condition must be optimal for easy peel removal or else albedo adhering to the segments will require additional hand labor or lye treatment. Processes are batch, not continuous, have high mechanical maintenance, and must be loaded and unloaded by hand. Finally, quality improvement is questionable and less economical than the traditional steam or hot-water loosening described above, which can handle high volumes in a continuous manner without the need to score the surface (Chace et al., 1940). Hot-water processes also kill potentially harmful microorganisms on the fruit. Infusion processes may even contaminate the fruit tissue during the vacuum treatment, if the infusion is not sanitary or has been contaminated.

8.3.1.2 *Sectionizing and Packaging*

After peeling, fruit sections may be sliced mechanically by hand placing the fruit in a cup or on a spindle, aligning bottom to top. A multiple blade head assembly then slices the fruit, separating the segments, which then must go through a seed and defect removal process. Brown International Corporation and FMC FoodTech Citrus Systems Division had each built sectionizing machines in the sectionizing industry's heyday. The mechanical process has not been entirely satisfactory because the segments are sliced and retain parts of the segment membrane. This membrane may be bitter (grapefruit) and tough to chew, decreasing acceptability. There have also been attempts to remove the segment membrane from the section with pectinases, citing the difficulties of losses with traditional mechanical or chemical methods (Ben-Shalom et al., 1986). Little success with alternate methods has the result that peeled fruit is mostly sectioned by hand, which involves a person with a sectionizing knife removing each individual segment.

Once recovered, the sections are graded and placed in containers with syrup made with sweetened juice. Different syrups may be used, depending on the product and whether it is cold-filled into glass with a preservative or canned (Kimball, 1991). Dietetic products with synthetic sweeteners are also produced. Some cold-filled sections in glass with syrup and preservatives may contain segments of other fruit to appeal to consumers' taste and increase marketability. These products may contain citric acid, ascorbate, benzoate, and/or sorbate. Canned product is heated after sealing to an internal temperature of near 90°C to stabilize the enzymes, inactivate microorganisms, and tenderize the membrane, in the case of mechanically sliced segments (Hannan, 1981). The hot cans are cooled with water to ambient temperature for storage and distribution. A pressurized rotating pressure cooker has also been used to increase the heating rate of cans as well as provide more rapid cooling, decreasing the process time to produce higher quality canned sections (FMC FoodTech Citrus Systems

Division, Lakeland, FL). Larger quantities of sections also have been packed into drums and frozen for bulk distribution (Chen et al., 1984).

8.3.2 Fresh-Cut Citrus

Attempts to revive the citrus sectionizing industry have resulted in a marketing program to package peeled, whole fruit in shrink-film containers so that the consumer may see the fruit. This product takes advantage of the trend to lightly processed fruits and vegetables and convenience of prepackaged fresh items for individual consumption. An improvement in the method for preparation of sections has stated that enzyme infusion processes facilitate peel removal and produce firmer and cleaner sections with less labor (Berry et al., 1988). Since the segments may remain intact as a whole fruit, not separated into sections, the marketing gurus refer to this product as "fresh-cut" citrus. A viable objective of preparing fresh-cut citrus is to increase the market appeal and convenience for consumers, expanding the product volume for the citrus industry. To expedite this objective, some technical data has been compiled and is available (Pao et al., 1998b). That consumers prefer individually prepackaged fresh fruits and vegetables, such as salads, has been documented and may prove to expand sales of fresh-cut citrus (Fresh Trends, 1998). Fresh-cut citrus is presently a developing market; however, commercial operations, retailers, and consumers have accepted fresh-cut items such as onions, lettuce, salads, melons, carrots, and pineapple (IFPA, 1998). The major limitation to large-scale commercialization of fresh-cut citrus is the rate-limiting step of peeling the fruit and the labor required for manufacture, inspection, and packaging this product. There is also some concern about microbial contamination and food safety since the product is not heated.

REFERENCES

Adams, B. and Kirk, W. (1991). Process for enzyme peeling of fresh citrus fruit, U.S. Pat. 5,000,967.

Ando, T., Suzuki, T., Ishii, K., Omura, H., and Yamazaki, J. (1988). Apparatus for separating juice sacs of citrus fruits, U.S. Pat. 4,738,194.

Anon. (1947). Juicells—New form of canned grapefruit, *Food Ind.*, Jan. 101–103.

Baker, R. A. and Wicker, L. (1996). Current and potential applications of enzyme infusion in the food industry, *Trends Food Sci. Technol.* **7**:279–284.

Ben-Shalom, N., Levi, A., and Pinto, R. (1986). Pectolytic enzyme studies for peeling of grapefruit segment membrane, *J. Food Sci.* **51**(2):421–423.

Berry, R. E., Baker, R. A., and Bruemmer, J. H. (1988). "Enzyme separated sections: A new lightly processed citrus product." In *Proc. Sixth Int. Citrus Congress*, R. Goren and K. Mendel, Eds., Balaban, Philadelphia, pp. 1711–1716.

Blake, J. R. (1980). Food product containing orange citrus vesicle solids, U.S. Pat. 4,205,093.

Braddock, R. J. (1983). Utilization of citrus juice vesicle and peel fiber, *Food Technol.* **37**(12):85–87.

Braddock, R. J. and Graumlich, T. R. (1981). Composition of fiber from citrus peel, membranes, juice vesicles and seeds, *Lebensm. Wiss. u. Technol.* **14**:229–231.

Bruemmer, J. H. (1981). Method of preparing citrus fruit sections with fresh fruit flavor ʾand appearance, U.S. Pat. 4,284,651.

Bruemmer, J. H., Griffin, A. W., and Onayemi, O. (1978). Sectionizing grapefruit by enzyme digestion, *Proc. Fla. State Hort. Soc.* **91**:112–114.

Chace, E. M., von Loesecke, H. W., and Heid, J. L. (1940). Citrus fruit products, USDA Circular No. 577, Washington, D.C.

Chen, C. S., Ting, S. V., and Hill, E. C. (1984). Predicting temperature changes in freezing of citrus sections contained in steel drums, *Trans. ASAE* **27**(5):1604–1609.

Eisenhardt, Jr., W. A., Saleeb, F. Z., McKay, R. P., and Zeller, B. L. (1984). Method for producing dried citrus pulp, U.S. Pat. 4,477,481.

Elsbury, J. (1932). The preparation of candied peel, *Food Manufact.* **7**:237–239.

Feeney, R. D., Prosise, R. L., McGrady, J., Niehoff, R. L., and Volker, D. A. (1988). Process for making brownies containing cellulosic fiber, U.S. Pat. 4,774,099.

Ferguson, R. R. and Fox, K. I. (1978). Dietary citrus fibers, *Trans. Citrus Eng. Conf.* **24**:23–33.

Fresh Trends (1998). Fresh-cut, *Packer*, **104**(54):40–50.

Hannan, Jr., H. (1981). Processing of grapefruit segments containing membranes, U.S. Pat. 4,254,157.

Hannigan, K. J. (1982). Dried citrus juice sacs add moisture to food products, *Food Eng.* **54**(3):88–89.

Hood, S. C. and Russell, G. A. (1916). The production of sweet orange oil and a new machine for peeling citrus fruits, USDA Bull. No. 399. Washington, D.C.

IFPA (1998). www.fresh-cuts.org.

Ifuku, Y., Uchiyama, H., and Hayashi, M. (1981). Method of separating and taking out pulp from citrus fruit, U.S. Pat. 4,294,861.

Kesterson, J. W. and Braddock, R. J. (1973). Processing and potential uses for dried juice sacs, *Food Technol.* **27**(2):50–54.

Kesterson, J. W. and Braddock, R. J. (1976). By-products and specialty products of Florida citrus, Fla. Agr. Exp. Sta. Tech. Bull. No. 784, University of Florida, Gainesville, FL, p. 87.

Kimball, D. A. (1991). *Citrus Processing Quality Control and Technology*, Chapman & Hall, New York.

Kock, R. W., Gosselin, P. G., and Reiboldt, H. N. (1990). Fluid impingement method & apparatus for fruit meat, U.S. Pat. 4,977,826.

Kock, R. W., Gosselin, P. G., and Reiboldt, H. N. (1991). Fluid impingement method for fruit extracting, U.S. Pat. 5,064,671.

Kolodesh, M. S., Toms, D., and Pierson, B. A. (1989a). Method of and apparatus for separating juice sacs from the sectional membranes of a fruit meat section, U.S. Pat. 4,873,106.

Kolodesh, M. S., Cash, Jr., W., Davis, J. E., Gosselin, P. G., Kock, R. W., Pierson, B. A., Reiboldt, H. N., Sabatelli, D. A., and Toms, D. (1989b). Method of and apparatus for extracting juice and meat from a fruit, U.S. Pat. 4,885,182.

LeGrand, F. and Stevens, N. C. (1991). Apparatus for peeling food products, U.S. Pat. 5,046,411.

Lynn, C. C. (1980). Citrus fiber additive product and process for making same, U.S. Pat. 4,225,628.

Mills, S. H. and Tarr, R. E. (1992). Fruit juice plus citrus fiber from pulp, U.S. Pat. 5,162,128.

Pao, S. C., Petracek, P. D., and Brown, G. E. (1996). Nonenzymatic fruit peeling method, U.S. Pat. 5,560,951.

Pao, S. C., Petracek, P. D., and Brown, G. E. (1998a). Citrus fruit peeling method, U.S. Pat. 5,817,360.

Pao, S. C., Petracek, P. D., and Brown, G. E. (1998b). Manual for fresh-cut citrus, FL Dept. of Citrus–Fresh Fruit, Lake Alfred, FL.

Passy, N. and Mannheim, C. H. (1983). The dehydration, shelf life and potential uses of citrus pulps, *J. Food Eng.* **2**:19–34.

Porzio, M. A. and Blake, J. R. (1983). "Washed orange pulp: Characterization and properties." In *Unconventional Sources of Dietary Fiber*, ACS Symp. Ser. 214, I. Furda, Ed., American Chemical Society, Washington, D.C., Chap. 14, pp. 191–204.

Shomer, I., Ben-Gera, I., and Fahn, A. (1975). Epicuticular wax on the juice sacs of citrus fruits: A possible adhesive in the fruit segments, *J. Food Sci.* **40**:925–930.

Ting, S. V. (1970). Alcohol-insoluble constituents of juice vesicles of citrus fruit, *J. Food Sci.* **35**:757–761.

Ting, S. V. and Rouseff, R. L. (1983). "Dietary fiber from citrus wastes: Characterization." In *Unconventional Souces of Dietary Fiber*, ACS Symp. Ser. 214, I. Furda, Ed., American Chemical Society, Washington, D.C., Chap. 15, pp. 205–219.

Tisserat, B., Galletta, P. D., and Jones, D. (1989). Induction of adventitious branches from cultured citrus juice vesicles—a potential means of proliferation, *Am. J. Bot.* **76**(12):1750–1758.

USDA (1962). *Chemistry and Technology of Citrus, Citrus Products, and By-products*, Agriculture Handbook No. 98, USDA, Washington, D.C.

Vincent, D. B., Vincent, D. A., and Roberts G. C. (1972). Citrus fruit peeling machine, U.S. Pat. 3,700,017.

Webb, J. D. and Houghton, A. F. (1976). Fruit peeling apparatus, U.S. Pat. 3,982,482.

Watanabe, S. (1985). Citrus fruit processing and food product, U.S. Pat. 4,560,572.

Watanabe, H., Hagura, Y., Ishikawa, M., and Sakai, Y. (1987). Cryogenic separation of citrus fruit into individual juice sacs, *J. Food Process Eng.* **9**:221–229.

Yoshida, Y. and Ueda, M. (1984). Citrus juice waste as a potential source of dietary fiber, *J. Jpn. Soc. Hort. Sci.* **53**(3):354–361.

CHAPTER 9

PEEL FIBER, CLOUD, AND PRODUCTS

Citrus fruit residue remaining from juice extraction amounts to approximately half the wet mass of the whole fruit, with the peel (flavedo and albedo) almost one-fourth the whole fruit mass (Tables 3.1–3.4). Products from the edible juicy portion of the fruit have considerable economic impact as foods for human consumption; however, there are also important uses of the peel as food ingredients. Compared with the edible portion of the fruit, for most human food applications, citrus peel must be processed free of bitterness, microbes, and other impurities. The traditional peel products of candied peel, marmalade, beverage bases, and purees have potential to use only small quantities of the total mass of peel residue, as they are mostly composed of sugar. Thus, citrus processors are seeking ways to utilize peel into value-added food products, which can increase profits and help with the considerable problem of waste disposal. The problem is particularly acute for smaller processors in countries whose industry is too small to support the investment needed to process the residue into dried cattle feed. This chapter will consider issues dealing with alternatives to making dried citrus pulp cattle feed as well as describe traditional and new processes for peel products for human food use.

9.1 PEEL FIBER

Citrus peel residue has a number of potential uses, which depend on the functional properties of the raw material, as well as the intended food product application. In reality, the functionality of the material is not complicated and depends on the basic peel composition. Whether or not citrus peel may be a valuable dietary fiber component in the human diet is beyond arguments made in this book, although this has been a topic of considerable discussion.

9.1.1   Composition

Chemical composition of citrus fruit components has been the subject of hundreds of research studies, so the interested reader should find it simple to search the literature for details not listed here. Compositional data useful for those interested in processing technology can be found summarized in textbooks of several authors (Braverman, 1949; Kefford and Chandler, 1970; Nagy et al., 1977; Sinclair, 1961, 1972, 1984; Ting and Rouseff, 1986). Like most fruits after water removal, soluble and insoluble carbohydrates are the major constituents of citrus peel. The water-soluble fraction contains glucose, fructose, sucrose, and some xylose, while pectin, hemicellulose, and cellulose constitute between 50 and 70% of the insoluble fraction (Ting and Deszyck, 1961; Ting and Rouseff, 1983). Considering the residue remaining after pectin recovery from orange peel, the dried fiber material was characterized to have 600 g/kg dietary fiber, half of which was cellulose (Aravantinos-Zafiris et al., 1994). Others have reported a decrease from 686 to 586 g/kg of total dietary fiber in peel as grapefruit matured (Larrauri et al., 1997).

When the nutritional importance of fiber in the human diet became a popular media item, considerable interest developed in citrus peel use as a fiber source. Since citrus peel has limited lignin, the so-called dietary fiber content is mostly represented by the insoluble carbohydrate fractions. Citrus pulp has been considered a source of insoluble dietary fiber that ranks with cereal bran in importance (McCready, 1977). For analysis, these fractions are generally isolated by extraction procedures with increasing concentrations of acid or base solutions, after alcohol/water extraction of the soluble sugars. The difficulty of obtaining clearly distinct fractions of pectin, hemicellulose, and cellulose by these methods has been discussed in a study describing these components in orange peel (Eaks and Sinclair, 1980). They stated that heterogeneity of the composition of pectin molecules contributes to the reason pure molecules of hemicellulose and cellulose cannot be isolated. Composition of Valencia peel was approximately as follows (percent dry peel weight): pectin (20%), hemicellulose (5–11%), cellulose–lignin fraction (9–16%), protein (4%), ash (3%). Orange peel albedo had higher pectin content in the dietary fiber fractions than a number of common fruits and vegetables, and it was suggested that the detergent methods underestimate pectin (Belo and de Lumen, 1981).

There has been some question about the suitability of using citrus peel in human foods because the peel may contain pesticide residues and other potentially toxic compounds (Mannapperuma, 1996). There have been no tolerance violations; however, 46% of grapefruits and 58% of oranges sampled contained residues of 6–11 different pesticides, with postharvest fungicides most frequently detected (USDA, 1994). Although, as of this writing, the FDA has passed no regulations pertaining to these substances, certain naturally occurring bioflavonoids in citrus peel, such as rutin, quercetin, and kaempferol glycosides, have been reported to show mutagenic activity in rats and in certain microbial tests. Subsequently, the Federation of American Societies for Experimental Biology reported to the FDA that unknown concentrations of quercetin and kaempferol glycosides in citrus peel makes estimation of dietary intake impossible and suggested additional studies (Anon., 1981). An extensive list and chemical properties of the common flavonoids in the various citrus fruit peels is available (Horowitz and Gentili, 1977). Also, the polymethoxylated flavones, such as tangeretin, nobiletin, and sinensetin, are enriched in the flavedo over other components of the fruit (Rouseff and Ting, 1979).

9.1.2 Processing

The masses and composition of fruit parts have enough differences to result in variability of the dietary fiber content, once a definition is established. One classification method generally describes water-soluble carbohydrates (glucose, fructose, starch, pectin, etc.) as "available carbohydrates" and pectin, hemicellulose, cellulose, and lignin as "unavailable carbohydrates," although some of the latter may be degraded by digestive microflora (Southgate, 1969). A current definition of water-soluble fibers includes pectin, gums, and mucilage, and insoluble ones as cellulose, hemicellulose, and lignin (Slavin, 1987). Soluble fibers lower blood cholesterol and may help reduce risk of heart disease. Insoluble fibers affect food transit time through the intestine, may be fermented by microbes, and are thought to reduce the risk of some cancers. Using Southgate's method, pectin is represented analytically from uronic acid analysis, hemicellulose from pentoses in a 5% H_2SO_4 extract, and cellulose is determined from hexose analysis of a 72% H_2SO_4 extract. The extract residue represents lignin and the sum is the total dietary fiber.

Recovery of each component part of the juice extraction residue would be required to manufacture fiber from each component. Dietary fiber values based on the Southgate scheme for the components of oranges and grapefruits have been determined and are listed in Table 9.1 (Braddock and Crandall, 1981; Braddock and Graumlich, 1981). As reported, the pectin and insoluble polysaccharide fractions are more concentrated than the lignin, which is important for both functional food uses and consideration as a dietary fiber. Because of the hull, the seed has significant amounts of lignin and cellulose, valuable fiber components. These values compare favorably with the dietary fiber content of other subtropical fruits and vegetables (Lund and Smoot, 1982), many of which are not readily available as processed products. However, as a dietary fiber source, the lignin component of citrus fruit is not comparable to that of the cereal grains.

Processing the juice extractor peel residue into flour for incorporation into foods has received commercial consideration. A general process is quite simple

TABLE 9.1 Dietary Fiber Content of Component Parts of Oranges and Grapefruits

Dietary Fiber	Orange				Grapefruit			
	Peel	Membrane	Juice Sac	Seed	Peel	Membrane	Juice Sac	Seed
Pectin	3.7	4.0	4.8	3.0	4.8	4.0	4.9	4.0
Hemicellulose	1.8	1.5	1.7	1.6	1.4	1.1	1.5	1.6
Cellulose	3.8	3.5	3.5	6.8	2.6	2.3	1.8	7.0
Lignin	0.3	0.7	1.0	3.2	1.2	0.7	0.9	3.3
Total	9.6	9.7	11.0	14.6	10.0	8.1	9.1	15.9

g/100 g Fresh Component[a]

[a]Moisture content: peel (flavedo + albedo) (70%), membrane (82%), juice sacs (85%), seeds (50%).

and usually involves some portion of the following steps: size reduction by hammermill; water (solvent) extraction to reduce soluble sugars, bitterness, and flavor; pressing for partial dewatering; dehydration; grinding or milling for further particle size reduction to flour; and packaging and storage under conditions to prevent off-flavors, mold, or insect growth. While some marketing claims relate the product to dietary fiber, the real intent has been to impart some functional property to the food and utilize the residue as a higher value product. One of the first products, citrus flour, was a controlled blend of orange and grapefruit peel, seeds and juice pulp, milled with sesame flour. This pale yellow product was produced, recognizing the need for careful blending from the variable fruit supply throughout the growing season, to have specific ratios from the different fruits for product uniformity (Belshaw, 1978). The manufacturing process involved chopping the residue into small pieces, reacting with calcium hydroxide, pressing to remove soluble sugars and water, blending with sesame flour, dehydration to <10% moisture, and milling to an acceptable small particle size (Lynn, 1980).

Recognizing that the albedo contains most of the insoluble carbohydrates, a method for recovery of albedo from juice extractor residue was proposed (Gerow, 1980). The procedure involved comminuting and milling peel in a water slurry and separation by screening, based on the larger particle size of the flavedo portion. Milling and grinding of these materials must be done carefully to avoid degradation of the physical structure of the fiber, decreasing water-holding capacity (Auffret et al., 1994). Another process recovered the albedo from the shaved peel from a Brown juice extractor for use as a water-binding agent. This involved coarse grinding of the albedo, extraction with water to remove sugars, alcohol (isopropanol) extraction to decolorize and debitter, and pressing or centrifugation for dewatering, followed by vacuum drying and milling to a flour (Altomare et al., 1985). Whole orange peels were the starting material in a very similar extraction-drying fiber recovery process stream, with other by-products recovered including sugar syrup, essential oils, and bioflavonoids (Bonnell, 1985).

Citrus peel products have also been prepared to take advantage of the gelling and functional properties of their pectin content. One process utilized sodium carbonate to adjust peel pH to 7–8.5, which allowed efficient demethoxylation by pectinesterase and reaction of the peel with its natural calcium content. The substrate was washed, pressed, and dried for use as a thickening agent in intermediate moisture foods and pet foods (Buckley and Mitchell, 1976; Mitchell et al., 1976). Use of acidification to solubilize and soften citrus peel, followed by grinding the mixture to puree, and reduce the particle size to <200 μm produced a binding agent product useful in food products (Ehrlich, 1996). A special milling process that involved hot carrier gas product flow and classification was used to prepare a dry flour product, consisting of specific proportions of peel, membrane, and seeds. It was claimed to produce a product in the 30- to 50-μm range that could be suspended in juice-based drinks (Tarr et al., 1991).

9.1.2.1 Utilization   The author has studied the composition of peel fiber and reviewed some proposed food applications related to its functional properties (Braddock, 1983; Braddock and Cadwallader, 1992). Dried peel or juice sac flour products have some major limitations for use in food products and proposed food processing applications. The dried products must be rendered flavorless and microbially stable during processing and then maintained in that state during storage and in the food product. The difficulty of these achievements have been documented in a study that showed the benefits of drying aids on sensory and vitamin C retention in dehydrated comminuted oranges (Vergara et al., 1995). A number of molds and spores shown to survive the heat of peel dehydration and pelleting may be readily isolated (Parish and Higgins, 1989). This could have severe consequences for food products where this material was added. For bulk commercial storage of dried fiber, sanitary practices must be followed during and after manufacture. At least one instance of warehoused citrus peel fiber contaminated with eggs and adults of the almond moth, *Ephestia cautella*, has occurred. Finally, the most common application appears to be as a functional food ingredient for the purpose of imparting textural properties associated with water binding and retention. This application must be carefully tested for stability in the product intended, as bitterness, haylike flavors, molds, and unexpected toughness or product shrinking may develop after time. Use as a true dietary fiber may have more marketing advantages than dietary functionality, especially in competition with fiber sources from grains. Other fiber sources may also be cheaper for the same functionality, as citrus peel fiber has high dewatering, solvent recovery, and preparation cost.

As discussed for dried juice sacs in the previous chapter, it should be noted for baking applications substituting citrus flour for grain sources, that the major component is pectin. Typical of such applications is the use of citrus flour as a bulking agent in low-calorie cake preparation (Glicksman et al., 1985) and brownies (Feeney et al., 1988). Advantages of using these crude peel flour preparations compared with purified pectin were claimed to be of better gel structures and cost advantage (Mitchell et al., 1979). Use of citrus peel in juice and beverages has also been studied. One patent proposed an extract of roasted citrus peel as a coffee or tea substitute (Ehrlich and Hess, 1985). A citrus juice product with a high dietary fiber content (40–80%) contained up to 4% citrus fiber and claimed no off-flavors and low viscosity (Mills and Tarr, 1992).

In a role as a dietary fiber source, citrus fiber, particularly pectin, has been reported to serve some useful dietary function. Pectin has been shown to bind with low-density serum lipoproteins and cholesterol, effectively reducing their levels in vitro (Baig and Cerda, 1983). Evidence is also available that dietary pectin lowers cholesterol by binding bile salts and forming gels to decrease intestinal absorption and may affect carbohydrate metabolism in mammals (Chang, 1983). Citrus pectin and fiber were the subject of a review documenting many references detailing dietary activity related to cancer, mineral nutrition, carbohydrate metabolism, and cholesterol level (Baker, 1994).

9.2 CLOUD

Many fruit juices and beverages are turbid, or cloudy, instead of transparent. Of these, consumers recognize pure citrus juices and citrus-flavored beverages as cloudy products, with clarity considered a quality defect. Cloudiness is due to suspension of the insoluble essential oils, pectins, lipids, and proteins in a finely divided matrix, which under proper conditions will not precipitate or rise to the top of the product. Pure juice cloudiness and its stability has been discussed in Chapters 5 and 6. This section will deal with processes for manufacturing beverage clouding agents as natural by-products from citrus peel residues. The impetus for such process development relates to the need for a natural replacement for brominated vegetable oils (BVO) to adjust specific gravities, enabling suspension of the cloud–flavor complexes in cloudy beverages. Because of suspected toxicity, BVO use has been prevented or limited in many countries.

Without BVO, cloudy beverages must contain added substances to maintain the colloidal suspension causing the cloudiness and flavor. These materials are stabilized first, in a concentrated emulsion, which must remain stable until, second, its dilution and stability in the final beverage. The complexity of the several systems involved, the cloud concentrate, the beverage with sugar or diet, CO_2 or not, long shelf life of beverages, and so forth, must be understood to realize the difficulty of producing a natural clouding agent from citrus peel. In spite of these problems, there are commercially available citrus clouds, made generally by pectic enzyme treatment and subsequent extraction, filtration, and concentration. Most of these natural citrus clouds have stability problems but serve functionally, without use of BVO in specific environments.

9.2.1 Composition

Citrus juice cloud is the model for preparation of the peel clouds. The value of preserving cloudiness of orange juice was recognized very early, and studies to identify the components responsible for cloud and its stability were performed. Valencia juice cloud was recovered by centrifugation, the insoluble classified components analyzed, and the composition determined as (%) ash (2.7), P_2O_5 (1.4), nitrogen (7.7), pectin (81.6), hemicellulose (2.8), and cellulose (2.6). The insoluble fraction/lipid ratio was 3/1, with peel oil found to combine with the lipid fraction. The cloud was similar for different fruit varieties and originated in the juice sacs rather than from mechanical disintegration of other fruit tissue (Scott et al., 1965).

Citrus beverage flavor and cloud emulsions contain water, stabilizers (usually hydrocolloids such as gum arabic or modified starch), organic acids, weighting agents (e.g., BVO or ester gum), and the flavoring oil. The composition of the peel extract clouding agents has not been definitively reported; but the material does function similarly to the artificial emulsions. A number of authors have described portions of the functional compounds, which seem to make these citrus clouds work in a beverage environment (Baker and Cameron, 1999). There appears to be a link between pectin and hesperidin in stabilizing the cloud. Since hesperidin is insoluble, it will form a cloud in solutions of pectin, with stability between pH 2 and 7 (Ben-Shalom et al., 1984). The stability of peel extract cloud has been related to oils, lipids, and pigments from the flavedo suspended with flavonoid crystals and pectin from the albedo (Shomer et al., 1985). Other studies implicate hesperidin as necessary for cloud stability of peel extracts, interacting with pectin (Ben-Shalom and Pinto, 1986).

The importance of protein in the stabilization of the colloidal substances forming the cloud has been suggested. A theory that pectin in peel extracts stabilizes cloud formed from protein coagulation by heat treatment in the range of pH 3.5–4.5 has been supported by evidence from enzymatic digestion of the pectin (Shomer, 1988). However, contrary evidence relates only a small portion of the cloud in juice to thermal processing effects and interaction between protein and pectin. Insoluble cellular protein is a very high percentage of the cloud in orange juice, which forms complexes with lower molecular weight constituents and ionic properties of the juice environment (Klavons et al., 1991). Instability of peel clouds is at least related to higher concentrations of sugars and citric acid, even in the absence of pectinesterase activity, as demonstrated for orange, grapefruit, and lemon juices (Rothschild and Karsenty, 1974).

9.2.2 Processing

Most of the processes for preparing a natural clouding agent from citrus peel residue are modeled after the patented process involving pectic enzyme treatment of the mixture to reduce viscosity, allowing concentration of the cloudy material to make a cloud base for beverage applications (Villadsen, 1968). Study of this process provided insight into the variables, which identified instability problems as well as reasons for the functionality of such cloudy materials. Key features of the process relate to the quality and variety of fruit, proper amount and type of activity of the enzyme preparation, extraction temperature, amount of water/peel, separation and particle size reduction of the colloidal cloud material, and stability of the concentrate and finished beverages (Larsen, 1969). A similar clouding agent was also prepared from citrus seed kernels (Kesterson and Hendrickson, 1969; Kesterson et al., 1972).

The process described by Larsen (1969) was studied in the author's laboratory and resulted in a similar flow diagram (Fig. 9.1) (Kesterson and Braddock, 1973). A limitation in our work was that adequate filters were unavailable to recover the proper particle size or remove nonessential soluble substances, which contribute to instability and viscosity. An interesting study showed that stable cloud particles near 1 μm could be recovered and concentrated with a microfiltration device to a higher concentration of cloudiness than by evaporation (Shomer and Merin, 1984). Current commercial peel cloud processes use a similar scheme, where production methods and use of specific equipment,

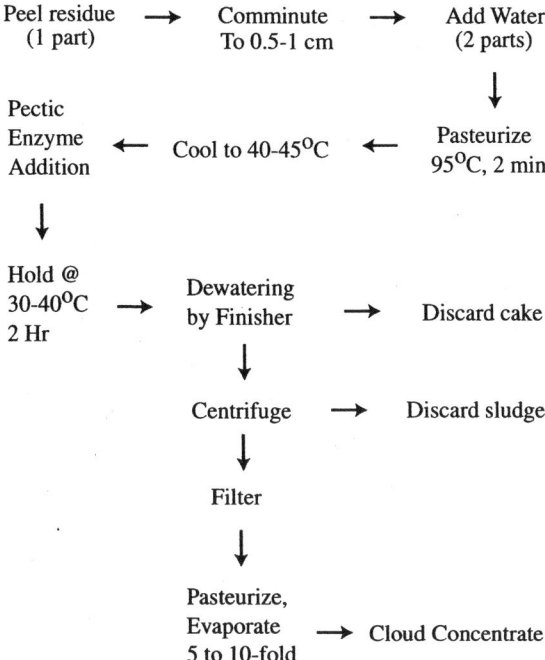

Figure 9.1 Typical process flow diagram for preparation of a natural cloud concentrate base from citrus peel.

such as decanter and clarifier centrifuges, were described (Bott and Schöttler, 1989).

Preparation of a stable peel cloud for the long life (>6 months at room temperature) required for beverages such as soft drinks depends on the critical parameters of oil content, final particle size, and rapid inactivation of the pectinesterase in the peel prior to addition of the commercial pectinase. Since there is no intention to add BVO to adjust specific gravity in beverages with peel cloud, the natural cloudy substances must have particles smaller than 1 μm to meet the stability criteria of Stoke's law and interactions with beverage constituents. Stability problems in the beverage are mainly due to ringing, flocculation, and coalescence, especially since the low beverage viscosity does little to aid in maintaining the cloud in suspension without the proper specific gravity. Even water hardness is important. These problems have been discussed in detail, with descriptions of the variables related to cloud stability (Tan, 1990).

9.2.3 Utilization

The importance of BVO for adjusting the specific gravity of beverage cloud emulsions can be illustrated by the evidence that a 12 °Brix solution with sugar and citric acid has a specific gravity of near 1.05. Thus, the critical citrus oil component (specific gravity, 0.85) in the oil phase of a stable emulsion should be close to that of the beverage. In reality, because of viscosity and ionic considerations, adjusting the specific gravity of the oil phase to 1.03 would be acceptable for orange drinks, and less (1.02) for lemon and lime drinks because the latter oils are more soluble (Oppenheimer, 1971). If sugar is absent, as in diet drinks, the specific gravity must be near that of water (<1.0) for cloud stability. The specific gravity, of course, is not adjusted in natural citrus peel clouds, so the method of preparing the emulsion (particle size, homogenization, etc.) and the composition and order of mixing the ingredients is very important (Tan, 1990).

There are a number of patents and proposed applications for making and utilizing citrus peel clouding agents. A peel shaver has been used to remove the flavedo and prepare the albedo fraction, which can be comminuted, finished, and pasteurized to make a cloudy mixture for citrus juice products (Douglas, 1972). A process for preparing a dried albedo clouding agent for use in dry beverage mixes involved enzyme treatment, washing with water and alcohol to remove off-flavors, and dehydration (Wiener and Haas, 1982). Continuous cold-pressed oil centrifuge discharges have also been evaluated as a raw material for preparation of beverage clouds. These substances provided good cloud, but had characteristic aromas and flavors, which could be imparted into the beverage (Bryan et al., 1973). Other research with citrus-oil-based cloud emulsions attempted to predict the degree of turbidity related to oil droplet size and degree of homogenization (Hernandez and Baker, 1991).

Addition of enzymes to citrus juices to facilitate manufacture of bases useful for producing cloudiness in beverages has been studied. One proposed product added pectic enzymes to destroy the pectin, allowing concentration to 80 °Brix to make a base for carbonated beverages (Atkins and Attaway, 1973). Addition of enzyme mixtures containing protease activity to citrus juice and concentrate aided in maintaining cloud stability in beverages and drinks to which the treated base was added (Wobben and Tan, 1983). Use of gums to help suspend pulp and insoluble matter in juice-containing drinks has also been proposed. One product is purported to assist pulp suspension in drinks without affecting the product's viscosity (Anon., 1983). Pectin plays an important role in cloud stability through viscosity adjustment of the beverage, substituting for weighting agents (Mezzino and Chuang, 1985; Klavons et al., 1994). An alternate approach considered fermentation of citrus residues to reduce the sugar and flavor content prior to extraction and cloud recovery (Sreenath et al., 1995). This process was unsuccessful.

9.3   OTHER PEEL PRODUCTS

9.3.1   Marmalade

Commercial processes for making citrus peel marmalade are many, and little has been added to the technology since the early published formulas. These generally require washing the fruit, extracting juice, mincing, slicing, or comminuting the peel, boiling to tenderize and remove bitterness, cooking with sugar (or syrup), portions of the juice, and pectin addition near the end of cooking (Chace, 1922; Chace et al., 1940; USDA, 1962). Sweet orange marmalade should be 30 parts fruit to 70 parts sugar. For the peel to be evenly distributed in the jelly, at least 0.5% high-grade pectin that sets rapidly, 65% sugar, and citric acid to adjust the pH to between 3.0 and 3.4 are requirements. A base peel may be prepared for use later in the season, when fruit is unavailable. This product may be chopped, washed, cooked, sulfited, and refrigerated for some months. For those planning to use a lot of citrus peel to make marmalade, consider that at 65 °Brix (sugar), the 30 parts raw fruit is only about 5% (dry matter) of the product mass. This means that a small amount of peel will make a lot of marmalade.

9.3.2 Candied Peel

Candied citrus peel is prepared mostly from orange or grapefruit peel. The primary use is in baking, where the candied peel is a condiment used for flavor, appearance, and texture in products like fruitcake. The peel of preference comes from reamers and can be used fresh or preserved with bisulfite. The half-peels are cooked to tenderize, remove bitterness, and undesirable flavors, drained, and diced. The diced peel is gradually equilibrated in sugar syrup to near 50 °Brix, then cooked in a 75 °Brix syrup to complete the process. Infusion of the sugars by vacuum accelerates the process (Elsbury, 1932). Food dyes (red, green, etc.) are added during the equilibration process to provide an attractive appearance in the baked product. The final candied product is drained by tumbling in a reel, allowed to air dry, and coated with a fine dust of cornstarch. It may be packaged in polyethylene containers for the individual consumer or larger packages for the commercial user.

A type of dried, candied citrus peel has been prepared by immersion in hot vegetable oil under vacuum to remove the water. The peel is then drained and impregnated with sugar to prepare an edible peel product (Swisher, 1975). It was found that the bitter flavor of candied grapefruit and orange peels complement the sweetness of certain confectionery products, much as does bitter chocolate. With this in mind, a number of recipes for candy products, fillings, chocolate clusters, fondants, and nougats containing candied orange or grapefruit peels has been published (Riedel, 1983).

9.3.3 Peel Seasoning

Citrus peel that has been carefully cleaned and dried at low temperatures to maintain the volatile oils is manufactured as a base for seasonings, marmalade, candied peel, condiments, and baked products. The process involves considerable hand labor and requires very fresh peel to produce a high-quality product from orange or tangerine peel. The peel is washed carefully and soaked in water through at least three rinses, with an antioxidant mixture (0.02% BHA, 0.05% ascorbic acid, 0.01% citric acid) the last rinse. This material is carefully dried to less than 5–10% moisture in a forced-air tray dryer. Maintaining the product temperature below 60°C (140°F) minimizes browning and loss of oils, preserving the flavor. Once the product is dried, it may be coarsely ground for packaging in opaque, airtight containers (under nitrogen) to retard oxidation and pigment loss.

9.3.4   Purees and Bases

A number of processes and fruit components are used to make purees and bases, including juice pulp, peel, and the whole fruit. Whole fruit purees contain the peel, oil, and seeds, as well as the juice components. Preparation generally involves grinding through a meat grinder, pasteurizing to inactivate pectinesterase, addition of pectinase to reduce viscosity, comminuting to make a paste or smooth puree, followed by sieving to remove broken seed hulls and non-disintegrated particles (Blakemore, 1962). Alternatively, whole fruits are washed and blanched, coarsely chopped, seeds separated, and the pulp is finely ground to a puree and homogenized (Cruse and Lime, 1970). Stable bases and purees for frozen desserts have been prepared utilizing comminuted whole oranges, gums, whipping and bodying agents, and vegetable oils (Blake et al., 1982). Other products are made by concentrating clarified juice to high °Brix, mixing it with pulp to make a high pulp base product (Chen and Chen, 1998).

High °Brix bases for beverages may be prepared by adding pectinases to pulpy juice to reduce viscosity, allowing concentration to 80 °Brix (Atkins and Attaway, 1973). Formulas for citrus juice bases to make drink or beverage bases may contain 65 °Brix concentrate or puree, gum arabic (0.05–0.1% in the finished beverage), food colorant, cold-pressed citrus oil and essence, citric acid, and water. The drink or beverage portion of the formula usually contains the sugar or sweetener. The final product may be 10 °Brix, 15–20 °Brix/acid ratio. A base for popsicles is similar, except that gelatin may be an added component. Since the base is usually held frozen to improve flavor, some preparations are not pasteurized, although microbial or enzymatic stability necessitates heat stabilization for long shelf life. Because they are ambient temperature products, the finished drinks usually contain a preservative, even when the base has been pasteurized.

A clear, bland, concentrated syrup base may also be prepared from the peel liquid by filtration of the press liquor. Syrup recovered from clarified, concentrated juice is too costly for most beverage applications since the competition may be much cheaper cane or beet sugar or high-fructose corn syrup. Such was the case for a liquid peel sugar product made from 20 °Brix concentrated press liquor by the following steps: prefiltration with traditional methods, hollow fiber ultrafiltration to yield a straw-colored permeate, cation, anion, and mixed bed resin columns to remove minerals, bitterness, and color, concentration by evaporation, and decolorizing with activated charcoal. The final product from this process was a water-clear, sweet, nonbitter 75 °Brix syrup from citrus fruit peel residue, which was composed mostly of fructose, glucose, sucrose, and trehalose (Breslau et al., 1976).

A countercurrent water extraction process has also been proposed for recovery of soluble solids from citrus peels and other fruit (Casimir, 1983). This process produces extracts with poor flavor (bitterness) and high viscosity.

REFERENCES

Altomare, R. E., Beale, R. J., Glicksman, M., Hegedus, E., Schulman, M., and Silverman, J. E. (1985). Citrus albedo bulking agent and process therefor, U.S. Pat. 4,526,794.

Anon. (1981). FASEB urges additional tests for citrus bioflavonoids, *Food Chem. News*, Jan. 12, pp. 15–16; June 29, p. 34.

Anon. (1983). New gum suspends fruit pulp without increasing viscosity, *Prepared Foods* **152**(3):154.

Aravantinos-Zafiris, G., Oreopoulou, V., Tzia, C., and Thomopoulos, C. D. (1994). Fibre fraction from orange peel residues after pectin extraction, *Lebensm.-Wiss. u.-Technol.* **27**(5):468–471.

Atkins, C. D. and Attaway, J. A. (1973). Natural orange base, U.S. Pat. 3,711,294.

Auffret, A., Ralet, M. C., Guillon, F., Barry, J. L., and Thibault, J. F. (1994). Effect of grinding and experimental conditions on the measurement of hydration properties of dietary fibres, *Lebensm.-Wiss. u.-Technol.* **27**(2):166–172.

Baig, M. M. and Cerda, J. J. (1983). "Citrus pectic polysaccharrides—Their in vitro interaction with low density serum lipoproteins." In *Unconventional Sources of Dietary Fiber*, I. Furda, Ed., ACS Symp. Ser. 214, American Chemical Society, Washington, D.C., pp. 185–190.

Baker, R. A. (1994). Potential dietary benefits of citrus pectin and fiber, *Food Technol.* **48**(11):133–136, 138–139.

Baker, R. A. and Cameron, R. G. (1999). Clouds of citrus juices and juice drinks, *Food Technol.* **53**(1):64–69.

Belo, Jr., P. S. and de Lumen, B. O. (1981). Pectic substance content of detergent extracted dietary fibers, *J. Agr. Food Chem.* **29**(2):370–373.

Belshaw, F. (1978). Citrus flour—a new fiber, nutrient source, *Food Prod. Devel.* **12**(7):36.

Ben-Shalom, N. and Pinto, R. (1986). The role of hesperidin in the turbidity of orange albedo aqueous extract, *Lebensm.-Wiss. u.-Technol.* **19**(2):158–160.

Ben-Shalom, N., Shomer, I., and Kanner, J. (1984). Pectin-hesperidin interaction in a citrus cloud model system: The effect of pectin degradation, *Lebensm.-Wiss. u.-Technol.* **17**(3):125–128.

Blake, J. R., Knutson, R. K., and Strommer, D. L. (1982). Composition for aerated frozen desserts containing fruit puree and method of preparation, U.S. Pat. 4,335,155.

Blakemore, S. M. (1962). Puree and method of making same, U.S. Pat. 3,031,307.

Bonnell, J. M. (1985). Process for the production of useful products from orange peel, U.S. Pat. 4,497,838.

Bott, E. W. and Schöttler, P. (1989). "Natural cloudy peel products." In *Centrifuges, Decanters and Processing Lines for the Citrus Industry*, Tech. Doc. No. 14. Westfalia Separator AG, Oelde, Germany, Chap. 4.5, pp. 22–25.

Braddock, R. J. (1983). Utilization of citrus juice vesicle and peel fiber, *Food Technol.* **37**(12):85–87.

Braddock, R. J. and Cadwallader, K. R. (1992). Citrus by-products manufacture for food use, *Food Technol.* **46**(2):105–110.

Braddock, R. J. and Crandall, P. G. (1981). Carbohydrate fiber from orange albedo, *J. Food Sci.* **46**(2):650–651, 654.

Braddock, R. J. and Graumlich, T. R. (1981). Composition of fiber from citrus peel, membranes, juice vesicles and seeds, *Lebensm.-Wiss. u.-Technol.* **14**:229–231.

Braverman, J. B. S. (1949). *Citrus Products; Chemical Composition and Chemical Technology*, Interscience, New York.

Breslau, B. F., Pensenstadler, D. F., and Mitchell, W. G. (1976). The recovery of sugar from citrus press liquor by ultrafiltration, *Trans. Citrus Eng. Conf.* **22**:53–74.

Bryan, W. L., Bissett, O. W., Wagner, C. J., and Berry, R. E. (1973). Potential by-products from waste citrus peel emulsion, *Proc. Fla. State Hort. Soc.* **86**:275–280.

Buckley, K. and Mitchell, J. R. (1976). Pectate gelled food products and method, U.S. Pat. 3,973,051.

Casimir D. J. (1983). Counter-current extraction of soluble solids from foods, *CSIRO Food Res. Quarterly.* **53**:38.

Chace, E. M. (1922). By-products from citrus fruit, USDA Circular No. 232, Washington, D.C.

Chace, E. M., von Loesecke, H. W., and Heid, J. L. (1940). Citrus fruit products, USDA Circular No. 577, Washington, D.C.

Chang, M. L. W. (1983). "Dietary pectin: Effect on metabolic processes in rats." In *Unconventional Sources of Dietary Fiber*, I. Furda, Ed., ACS Symp. Ser. 214. American Chemical Society, Washington, D.C., Chap. 11, pp. 185–190.

Chen, C. S. and Chen, W. A. (1998). Method for producing ready to pour frozen concentrated clarified fruit juice, fruit juice produced therefrom, and high solids fruit product, U.S. Pat. 5,756,141.

Cruse, R. R. and Lime, B. J. (1970). How to make whole citrus fruit puree, *Food Eng.* **42**(6):109–112, 115, 116.

Douglas, P. L. (1972). Cloud fortified citrus fruit juices, U.S. Pat. 3,647,475.

Eaks, I. L. and Sinclair, W. B. (1980). Cellulose-hemicellulose fractions in the alcohol-insoluble solids of Valencia orange peel, *J. Food Sci.* **45**(4):985–988.

Ehrlich, R. M. (1996). Pecto-cellulosic product from whole citrus peel and other materials, U.S. Pat. 5,567,462.

Ehrlich, J. R. and Hess, H. (1985). Beverages obtained from alcoholic treatment of roasted citrus fruit peels, U.S. Pat. 4,497,842.

Elsbury, J. (1932). The preparation of candied peel, *Food Manufact.* **7**:237–239.

Feeney, R. D., Prosise, R. L., McGrady, J., Niehoff, R. L., and Volker, D. A. (1988). Process for making brownies containing cellulosic fiber, U.S. Pat. 4,774,099.

Gerow, G. P. (1980). Separating citrus peel into albedo and flavedo components, U.S. Pat. 4,225,625.

Glicksman, M., Frost, J. R., Silverman, J. E., and Hegedus, E. (1985). Process for preparing a high quality, reduced-calorie cake, U.S. Pat. 4,526,799.

Hernandez, E. and Baker, R. A. (1991). Turbidity of beverages with citrus oil clouding agent, *J. Food Sci.* **56**(4):1024–1026.

Horowitz, R. M. and Gentili, B. (1977). "Flavonoid constituents of citrus." In *Citrus Science and Technology*, Vol. 1, S. Nagy, P. E. Shaw, and M. K. Veldhuis, Eds., AVI, Westport, CT.

Kefford, J. F. and Chandler, B. V. (1970). *The Chemical Constituents of Citrus Fruits, Advances in Food Research*, Academic, New York.

Kesterson, J. W. and Braddock, R. J. (1973). Unpublished research records, 12/17/73, 6/19/74, 12/3/74.

Kesterson, J. W. and Hendrickson, R. (1969). Citrus seed clouding agent for beverage bases, *Am. Perfumer Cos.* **84**(4):37–39.

Kesterson, J. W., Hendrickson, R., and Atkins, C. D. (1972). Citrus seed clouding agent for beverage bases and food products, U.S. Pat. 3,660,105.

Klavons, J. A., Bennett, R. D., and Vannier, S. H. (1991). Nature of the protein constituent of commercial orange juice cloud, *J Agr. Food Chem.* **39**(9):1545–1548.

Klavons, J. A., Bennett, R. D., and Vannier, S. H. (1994). Clouding agent for beverages and method of making, U.S. Pat. 5,286,511.

Larrauri, J. A., Rupérez, P., Borroto, B., and Saura-Calixto, F. (1997). Seasonal changes in the composition and properties of a high dietary fibre powder from grapefruit peel, *J. Sci. Food Agric.* **74**:308–312.

Larsen S. (1969). Examination of a cloudy material prepared from citrus peel, *Fruchtsaftforschung Technol., IFU*, Vol. 9, Aarhus, Juris Druck, Zürich, pp. 109–133.

Lund, E. D. and Smoot, J. M. (1982). Dietary fiber content of some tropical fruits and vegetables, *J. Agr. Food Chem.* **30**(6):1123–1127.

Lynn, C. C. (1980). Citrus fiber additive product and process for making same, U.S. Pat. 4,225,628.

Mannapperuma, J. D. (1996). "Residual management in fruit processing plants." In *Processing Fruits: Science and Technology*, Vol. 1, *Biology, Principles, and Applications*, L. P. Somogyi, H. S. Ramaswamy, and Y. H. Hui, Eds., Technomic, Lancaster, PA, Chap. 15, pp. 461–499.

McCready, R. W. (1977). "Carbohydrates: Composition, distribution, significance." In *Citrus Science and Technology*, Vol. 1, S. Nagy, P. E. Shaw, and M. K. Veldhuis, Eds., AVI, Westport, CT.

Mezzino, J. F. and Chuang, L. Y. (1985). Pectin-based clouding agent, U.S. Pat. 4,529,613.

Mills, S. H. and Tarr, R. E. (1992). Fruit juice plus citrus fiber from pulp, U.S. Pat. 5,162,128.

Mitchell, J. R., Buckley, K., and Burrows, L. E. (1976). Gelling and thickening agents, U.S. Pat. 3,982,003.

Mitchell, J. R., Buckley, K., and Burrows, I. E. (1979). Food binding agent, U.S. Pat. 4,143,172.

Nagy, S., Shaw, P. E., and Veldhuis, M. K. (1977). *Citrus Science and Technology*, Vol. 1, AVI, Westport, CT.

Oppenheimer, A. (1971). Clouding agents for the food industry, *Food Product Devel.* **5**(3):90, 92, 94.

Parish, M. E. and Higgins, D. P. (1989). Yeasts and molds isolated from spoiling citrus products and by-products, *J. Food Protect.* **52**(4):261–263.

Riedel, H. R. (1983). Pralines made with candied orange or grapefruit peel, *Confectionary Production* **48**(10):526–528.

Rothschild, G. and Karsenty, A. (1974). Cloud loss during storage of pasteurized citrus juices and concentrates, *J. Food Sci.* **39**(5):1037–1041.

Rouseff, R. and Ting, S. V. (1979). "Analysis of polymethoxylatedflavones in orange juice and fruit parts." In *Liquid Chromatographic Analysis of Food and Beverages*, Vol. 2, Academic, New York, pp. 537–558.

Scott, W. C., Kew, T. J., and Veldhuis, M. K. (1965). Composition of orange juice cloud, *J. Food Sci.* **30**:833–837.

Shomer, I. (1988). Protein self-encapsulation: A mechanism involved with colloidal flocculation in citrus fruit extracts, *J. Sci. Food Agric.* **42**:55–66.

Shomer, I. and Merin, U. (1984). Recovery of citrus cloud from aqueous peel extract by microfiltration, *J. Food Sci.* **49**(4):991–994.

Shomer, I., Lindner, P., Vasiliver, R., Kanner, J., and Merin, U. (1985). Colloidal fractions of citrus fruit aqueous peel extract, *Lebensm.-Wiss. u.-Technol.* **18**(6);357–365.

Sinclair, W. B. (1961). *The Orange: Its Biochemistry and Physiology*, University of California Press, Berkeley, CA.

Sinclair, W. B. (1972). "Grapefruit by-products." In *The Grapefruit*, University of California Press, Riverside, CA, Chap. 6, pp. 502–551.

Sinclair, W. B. (1984). *The Biochemistry and Physiology of the Lemon and Other Citrus Fruits*, University of California Press, Oakland, CA.

Slavin, J. L. (1987). Dietary fiber: classification, chemical analyses, and food sources. *J. Am. Diet. Assoc.* **87**(9):1164–1171.

Southgate, D. A. T. (1969). Determination of carbohydrates in foods. I. Available carbohydrates. II. Unavailable carbohydrates, *J. Sci. Fd. Agric.* **20**:326–330, 331–335.

Sreenath, H. K., Crandall, P. G., and Baker, R. A. (1995). Utilization of citrus by-products and wastes as beverage clouding agents, *J. Ferment. Bioeng.* **80**(2):190–194.

Swisher, H. E. (1975). Dehydrated citrus peel product, U.S. Pat. 3,868,466.

Tan, C. T. (1990). "Beverage emulsions." In *Food Emulsions*, K. Larsson and S. E. Friberg, Eds., Marcel Dekker, New York, Chap. 10, pp. 445–478.

Tarr, R. E., Burkes, A. L., and Mills, S. H. (1991). Method for preparing ultrafine citrus fiber and derivative fiber-enriched citrus beverages, U.S. Pat. 5,073,397.

Ting, S. V. and Deszyck, E. J. (1961). The carbohydrates in the peel of oranges and grapefruit, *J. Food Sci.* **26**(2):146–152.

Ting, S. V. and Rouseff, R. L. (1983). "Dietary fiber from citrus wastes: Characterization." In *Unconventional Sources of Dietary Fiber*, ACS Symp. Ser. 214, I. Furda, Ed., American Chemical Society, Washington, D.C., Chap. 15, pp. 205–219.

Ting, S. V. and Rouseff, R. L. (1986). *Citrus Fruits and Their Products: Analysis and Technology*, Marcel Dekker, Inc., New York.

USDA. (1962). Chemistry and technology of citrus, citrus products and by-products. Agric. Hndbk. No. 98, USDA, Washington, D.C.

USDA (1994). Pesticide data program, Summary of 1992 data, Agricultural Marketing Program, April, Washington, D.C.

Vergara, F., Welti, J., and Barbosa-Cánovas, G. (1995). Stability of dehydrated comminuted orange products. Influence of maltodextrin and storage temperature, Adv. Food Eng. Proc. 4th Conf. Food Eng., G. Narsimhan, M. Okos, and S. Lombardo, Eds., Purdue Univ., West Lafayette, IN.

Villadsen, K. J. (1968). Preparation of clouding and coloring agent for soft drinks, U.S. Pat. 3,404,990.

Wiener, C. and Haas, G. J. (1982). Dried albedo clouding agent and process therefor, U.S. Pat. 4,335,143.

Wobben, H. J. and Tan, H. B. (1983). Process for the preparation of citrus juice containing beverages with improved cloud stability, U.S. Pat. 4,388,330.

CHAPTER 10

DRIED PULP, PELLETS, AND MOLASSES

Citrus fruit peel and membrane residue from the juice extractors is the primary waste fraction from processing, amounting to approximately 40 to 50% of the wet fruit mass. This material is conveyed from the juice extraction room to a peel bin at the plant's feed mill, or in some cases to a surge tank for mixing and pumping to the feed mill. The large amount of wet residue necessitates manufacturing a large volume product, such as cattle feed (Braddock and Cadwallader, 1992). This utilization provides the >2 million mt of dried citrus pellets available from the world's citrus processing plants. The final dried product is pelletized for economy of storage and handling. During the drying operation, some of the water and sugars are mechanically recovered by pressing the residue. This press liquid may be concentrated by evaporation to make a molasses, available for cattle feed, fermentation products, or adding back to the peel entering the dryer.

10.1 DRIED PULP AND PELLETS

The use of citrus peel residue as dried pulp for cattle feeding was suggested before 1920 (Walker and McDermitt, 1917) and nutritional studies were performed feeding wet and dried orange pulp to livestock in the 1920s (Mead and Guilbert, 1926; Scott, 1926). This was extended to grapefruits and other citrus with the production of commercial quantities in the 1930s (Neal et al., 1935; Arnold et al., 1941). A description of the processing technology included mass flow diagrams with and without pressing, drying variables, and a method for calculating the amounts of press cake and liquor based on the moisture content of the wet residue (Heid, 1945). Similar and other details of this process may be found in the publication by Kesterson and Braddock (1976) and other citations (Braverman, 1949; Sinclair, 1961, 1972; Timmons, 1950; USDA, 1962).

10.1.1 Processing

The importance of feed mills to operation of citrus processing plants has been described by listing three primary characteristics: (1) Feed mills are the largest energy consumers of a plant. (2) They are the primary pollution control point, especially for liquid wastes. (3) They produce the main by-products of cattle feed, molasses, d-limonene, and in some cases pectin peel and ethanol (Odio, 1993).

10.1.1.1 Handling Wet Residue Citrus plants have the capacity to accumulate large masses of extractor residue in closed peel bins. The traditional processes to move peel involve screw-and-slide conveyors, elevators, and trucks to move the peel residue, trash (leaves, stems, etc.), cull fruit, and juice pulp to the peel bins. From the peel bin, the residue is delivered out through manual slide gates into screw conveyors to the hammermills for shredding to reduce the particle size to approximately 0.6–2.0 cm (0.25–0.75 inch). The shredding operation optimally would reduce the peel to a narrow size range with the smallest size most efficient for water removal and the largest least efficient. Particles below 0.6 cm (0.25 inch) shrink significantly during drying and are difficult to separate from the dry process stream, creating dust and fines; and pieces larger than 2.0 cm (0.75 inch) do not dry uniformly. The fines from small particles may be lost in the stack emissions or burned in the dryer, contributing to yield reduction and potential air pollution. Large pieces may not attain dryness during residence time in the dryer and can mold or contribute to so-called spontaneous combustion in the stored product.

10.1.1.2 Liming and Pressing Addition of lime to the residue is a necessary processing aid, recognized in the 1930s, as it was used to aid in dehydration of beet pulp, clarification of cane juice during sugar manufacture, and for other pectinaceous materials. In modern citrus processing, it may be added several ways, as a powder, a water, or molasses slurry, either before or after the hammermill (or in both locations) (Fig. 10.1). Calcium oxide (CaO, called quicklime) is used, which readily hydrates with water (called slaking) in the residue and molasses, liberating heat and forming calcium hydroxide [$Ca(OH)_2$], which is quite basic (pH 12, saturated solution). Burning limestone in a kiln makes CaO. Attempts to use agricultural lime (limestone), which is calcium carbonate ($CaCO_3$), are not successful because of its limited solubility. CaO is cheap and readily available; however, MgO, where available, will also perform the same function and results in less evaporator scale (Hillis et al., 1968).

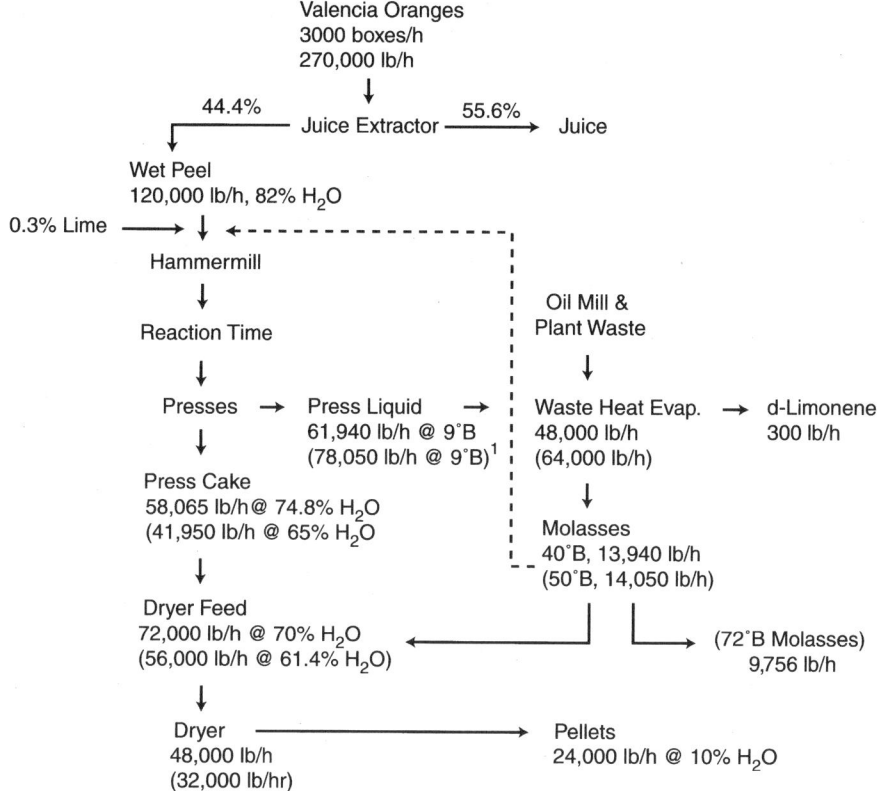

Figure 10.1 Flow and mass balance of feed mill processes for manufacturing dried citrus pulp pellets, molasses, and *d*-limonene from Valencia oranges.

Based on the wet weight of the residue, lime concentration required is between 0.2 and 0.5%. Lemons, limes, and grapefruits require more and oranges less lime in this range, depending on the residue acidity. Key to a successful liming operation is the proper reaction of the lime with the residue. This is a typical diffusion process, controlled by the size of the shredded peel and the reaction time. The traditional process involves mixing in a reaction screw, with the optimum time between addition and the presses in the range of 10–15 min. The initial wet residue has a pH between 3 and 4.5, locally rises as the lime reacts to greater than pH 9–10. At the high pH, the flavonoids in the peel react to produce a yellow color, which disappears when the pH drops back to neutrality. The press cake and press liquor remaining yellow is an indication that too much lime has been added. As equilibrium is achieved, the pH drops to the optimum (measured in the press liquor) of 6.5–7.

The chemical reaction of the lime with the residue is twofold: to neutralize the peel acidity and to de-esterify the pectin. The latter occurs at the higher pH, producing pectic acid, which reacts with the calcium in the lime to form the calcium pectate salt, liberating water and methanol during pressing. Buffer calculations of the amounts of lime used to react with the pectin and acids in the peel are based on the pK_a of citric acid ($pK_{a3} = 5.4$) and galacturonic acid ($pK_a = 5.4$) (pectin average molecular weight = 70,000). It can be calculated that 75% of the base is used to neutralize and hydrolyze the pectin to pectic acid because the amount in the residue is 5–10 times greater than the peel acids. This is the basis for substituting spent caustic for a portion of the lime used in the feed mill (Johnston, 1998), although calcium is still required to form the salt for water removal by pressing.

Continuous liming and mixing systems involving fluid pumping of the peel residue and molasses, rather than traditional reaction screws, is being implemented in citrus feed mills (Odio, 1993; Marques, 1995; Johnston, 1998). This process innovation eliminates the reaction conveyor and provides more efficient and timely reaction of lime with the residue by replacing the reaction screw with a tank, agitator, and slurry pumps. The process requires mixing the limed, shredded residue with press liquor and/or molasses in a system where it may be pumped from the reaction tank to the presses. Process claims include more homogeneous product and efficient lime reaction, lower press cake moisture, lower power consumption, investment, maintenance, and less labor cost.

When the properly limed and reacted residue is pressed in the continuous screw presses, the moisture content is reduced from the normal 78–82% to the 70–72% range. Residue from juice extraction with orange peel oil recovery, where water is added to recover the oil, is near the value of 82%; and peel without oil recovery or with minimal water absorbed will be near 78% moisture. By addition of molasses in the 40–50 °Brix range to the residue ahead of the presses, press cake moisture may be as low as 60–62% feeding the dryer (Odio, 1993). Press cake moisture reduction below the 70% value also may be achieved through a scheme of molasses addition and use of double pressing (Cook, 1983; Johnston, 1995). The advantage of reducing the press cake moisture to lower values is that the peel dryer has less water to evaporate, shifting the evaporation load to the waste heat evaporator, which has the benefit of multiple effect economy. This is illustrated by examination of Figure 10.1, which shows the advantage of the dryer and waste heat evaporation requirements for press cake at 74.8 or 65% moisture generating 40 or 50 °Brix molasses. Generally, the lower moisture (65%) uses less fuel and can dry about 50% more press cake than at the higher value. Values for grapefruit juice extractor residue are similar to those of Valencia oranges (Fig. 10.1); however, the initial masses of juice (47.9%) and residue (52.1%) are slightly different because of the thicker peel of grapefruits.

The pressing operation generally consists of feeding the limed residue to a horizontal or vertical continuous, slowly turning screw spindle mounted inside a reinforced screen, much like a juice finisher. The pressure is achieved against the screen with the turning screw by restricting the opening of the press cake discharge. The press liquor flows through the screen openings to a collection tank for evaporator feeding, while the press cake falls to a screw feeding the dryer inlet. If the residue has too little lime (pH <6) or short reaction time, pectin will not be degraded, the press cake will be slippery, will not press or dry properly, and the press liquor will tend to have high viscosity and foam. Too much lime may cause hardening of the cake, blinding of the presses, excessive dust in the dryer exhaust, evaporator scale and may affect the calcium nutrition of the dried pulp. Screening the press liquor to remove fine particles passing through the press screens aids in efficient evaporation to molasses.

10.1.1.3 Dehydration   Citrus pulp is manufactured by either steam-tube, direct-fired triple-pass and single-pass rotary dryers. Most peel dryers now in service are the single-pass type, constructed of a long, insulated, cylindrical rotating shell, with the furnace mounted parallel so that flame is not directed on the incoming cake or recycled vapor. After entering, the press cake tumbles in the hot vapor stream in the rotating drum, moving toward the exit as it dries by the hot gas flow through the dryer. Drying rate curves show a falling rate of drying for citrus pulp (Braddock and Miller, 1978). The mass flow of hot gasses containing water evaporated from the press cake is driven by a fan mounted above a cyclone separating the dried pieces from the vapors at the dryer exit. Up to 50% of the exhaust gas is recirculated through the drum to increase the vapor concentration close to the dew point (82°C) at the exhaust temperature, allowing efficient operation of the waste heat evaporator. The exit gas temperature of the dryers is near 150°C (300°F), but with care the product temperature should not exceed 75–80°C (165–175°F). Higher product temperatures decrease the nutritional value and palatability for cattle by burning some of the carbohydrates in the dried pulp (Ammerman et al., 1965).

Natural gas, fuel oil, or other materials, such as sugar cane bagasse, may fuel dryers. Occasionally, the value of dried pulp is very low, and interest in burning the peel as fuel increases in the industry. The high heat value (heat of combustion) for dried citrus pulp has been determined to be 4150 cal/g (7470 Btu/lb at 0% moisture) (Kesterson et al., 1979). Citrus pulp may be compared with current (1998, Florida) average fuel values for No. 6 fuel ($0.35/therm) and natural gas ($0.31/therm) (100,000 Btu/therm). At a value of $100.00/ton as cattle feed pellets, dried pulp value as fuel would be $0.67/therm (13.38 lb pulp/therm), neglecting the energy needed to dry to the most efficient combustion point (25% moisture), clearly showing the higher value as animal feed.

Drying technology improvements in use for other products have also been considered for citrus pulp application. Of these, automated process control of feed mill unit operations has been implemented and proven to increase efficiency and output of uniform quality product and lower energy costs. Process control may be as simple as maintaining constant firing rates for the peel dryer furnace (Cook, 1987) or require inputs from the peel bin, press liquor tank pH and level, dryer input and output moistures, temperatures, and the waste heat evaporator (Sykora, 1997). Drying peel with microwave energy might be useful as a quality control method, but drying times were too extensive to be useful on-line for continuous drying (Miller and Braddock, 1982). Steam drying of peel under conditions of superheating by way of heat exchangers has been proposed (Covington, 1983). This process allows evaporation of product moisture where the transport steam and dried pulp are separated in a cyclone, with advantages of less fuel and higher quality product. High fuel costs also increase interest in cogeneration of electricity by citrus processors (Coxe, 1987). One scheme involves the use of gas turbines generating electricity and heat for steam or for dehydrating peel in either a parallel or series arrangement with the dryer furnace (Leo, 1982). Energy to dry peel ranges from near 900 kcal/kg water evaporated (1600 Btu/lb), considering only the dryer, to 400 kcal/kg (700 Btu/lb) with dryer and waste heat evaporator (Braddock et al., 1977).

10.1.1.4 Pelleting and Cooling

The loose dried pulp may be marketed in the state it exits the dryer but more commonly is pelletized by extrusion through a die. Whether the residue is limed and dried without pressing or pressed with molasses added, the final pulp should exit the dryer at near 12–15% moisture, enter the pellet mill, and exit the pellet cooler below 12% moisture. Friction and steam in the pellet mill generally will lower the moisture content to near the range required, while ambient air pulled through the pellets in the cooler will reduce the moisture slightly. The final product, ideally, will be warm to the touch (not hot), light brown, and be between 8 and 12% moisture. The integrity of citrus pellets is generally satisfactory for short-time handling and shipping without use of bonding agents, as the sugars and molasses caramelize, hardening the pellets during extrusion through the die.

10.1.1.5 Storage and Properties

Pellets are bulk-stored in warehouses in piles on the floor or in concrete or steel silos at the citrus plant and embarkation ports. Fires or smoldering of stored product is a serious concern and usually has been related to rainwater leaks or product with >14–15% moisture contents mixed with properly dry material. Product may also burn in the dryer and be inadvertently mixed with pellets in the warehouse. Fires will usually smolder and burning will be slow in confined areas where oxygen is limited; however, violent burning can take place with sudden air input. In warehouses, smoldering pellets in a pile accessible by front-end loader may be removed. In silos, fires will either eventually burn out all the fuel or stop if air leaks can be sealed.

Moisture equilibrium of dried pulp during seasonal storage in the Florida climate (average relative humidity, 60%) was near 12% moisture (Ross and Kiker, 1967). To maintain pellet moisture below 12%, contact with air greater than 50–60% relative humidity at 25°C should be minimized. The value (12%) is in the safe range for not undergoing spontaneous combustion or supporting mold growth (Braddock and Miller, 1978). The potential of microbial spoilage of improperly dried or stored pellets is quite high since the feed mill and bulk storage environments are anything but sterile. Ten different molds and yeast

TABLE 10.1   Properties of Citrus Pellets Relating Mass, Area, and Volume

	Pellet Diameter	
Property	0.95 cm	0.64 cm
Pellets/kg	1400	1500
Area (cm^2/pellet)	4.6	4.5
Area (cm^2/kg)	6500	7000
Pellet volume (cm^3/kg)	1070	1010
Bulk density (kg/m^3)	800	825
Void space (%)	60	63
Specific gravity	0.7	0.74

have been isolated and identified from commercial citrus pellets (Parish and Higgins, 1989).

Citrus pellets are extruded in two diameters, 0.64 and 0.95 cm (one-fourth and three-eighths inch), with the 0.95-cm pellet more common. There are some larger pellets, 1.3 cm, manufactured, although larger pellets disintegrate easier. The average pellet length (0.65–1.5 cm) depends on breaking from the extruder and handling procedures. The bulk density of pellets is twice the value for the loose dried pulp and results in significant savings in storage and transportation costs. Some pellet properties are presented in Table 10.1. The bulk density of dried pulp is in the range of 400 kg/m^3 (20–25 lb/ft^3), while pellets are near 800 kg/m^3 (50 lb/ft^3). Particle size distributions indicated that citrus press cake and dried pulp had similar distributions, with the highest percentage in the 2- to 5-mm range (Braddock and Miller 1978).

10.1.2   Composition

10.1.2.1   *Nutritional Value*   Dried citrus pulp and pellets are highly palatable to cattle, high in digestible energy and fiber, and low in digestible protein. If fed at levels no higher than 40% of the ration, citrus pulp is considered equal in feeding value to ground, snapped corn and corn meal. Excessively dark or charred pellets are not as palatable, having less protein and digestible energy than good-quality product. The average percent digestibility for each of the nutrients of citrus pulp is protein 51%, fat 85%, fiber 68%, and nitrogen-free extract (NFE) 89%. The average total digestible nutrient (TDN) value for citrus pulp is about 75, with an estimated energy value for beef cattle of about 1.5 kcal/kg pulp (Chapman et al., 1972).

10.1.2.2   *Nutrient Composition*   The nutrient composition of the various components of citrus pulp cattle feed is listed in Tables 10.2 and 10.3. Compared on a dry solids (0% moisture) basis, the different products have similar

TABLE 10.2   Nutrient Composition of Citrus Residue and Feed Products

Product[a]	Moisture (%)	Protein (%)	Fat (%)	Fiber (%)	NFE (%)	Ash (%)
Whole orange	85.2	1.0	0.3	1.6	11.3	0.6
Whole grapefruit	86.4	1.0	0.6	1.4	10.0	0.5
Wet orange residue	81.5	1.2	1.8	2.2	12.5	0.7
Orange press cake	71.7	2.0	2.7	4.4	17.8	1.5
Grapefruit press cake	74.8	2.2	1.2	4.6	15.7	1.5
Dried pulp pellets	10.6	6.3	3.2	10.2	63.0	6.7
Dried grapefruit pulp	10.0	5.8	4.9	10.9	62.5	5.9
Ensiled orange peel[b]	84.0	1.3	—	2.6	13.0	0.9

[a] Values adapted from Kesterson and Braddock (1976).
[b] Values adapted from Ashbell and Donahaye (1986).

proximate analyses, with probably the largest differences in the ash fraction (Table 10.2). The mineral nutrients show that the calcium content varies from 0.8 to 2.1% and is the probable cause for the ash variability (Table 10.3). The major reason for calcium variability is the differing amounts of lime added during processing. Too much lime will increase the calcium/phosphorus nutritional ratio above the average, and the product may not meet specifications.

10.1.2.3 *Wet Residue, Press Cake, and Silage*   Use of the wet residue, press cake, and silage from the wet residue as cattle feed has been practiced since the 1920s. With these materials, studies have shown that cattle weight gain is directly proportional to the TDN intake (Chapman et al., 1972). The material typically contains too much water for cattle to eat enough, compared

TABLE 10.3   Mineral Nutrient Composition of Dried Citrus Pulp[a]

Mineral	Mean ± std dev		Range
Calcium (%)	1.4	0.27	0.8–2.1
Magnesium	0.1	0.01	0.1–0.2
Phosphorus	0.1	0.02	0.1–0.2
Potassium	1.1	0.10	0.8–1.3
Sodium	0.1	0.04	0.1–0.3
Sulfur	0.1	0.04	0.1–0.2
Iron (ppm)	98.7	41.8	31–214
Copper	6.2	2.9	3–17
Zinc	9.9	2.2	6–13
Manganese	5.7	0.6	4–7
Cobalt	0.1	0.02	0.04–0.1

[a] Values adapted from Kesterson and Braddock (1976).

to the dry products. Also, peel oil can contribute to unpalatability. Because of the presence of soluble carbohydrates in the residue, the peel will ferment and sour readily, before cattle can eat it. Attracting insects and drainage of liquid result in environmental problems. Hauling the residue, which is 80% water, is also expensive. Considering these issues as well as nutritional value, press cake has advantages over wet residue, and a bag system has been developed to handle and distribute it on a limited scale (Solis, 1997).

The fermentable nature of citrus residue has also resulted in attempts to ensile it for preservation (Braverman, 1949; Becker et al., 1954). After about 3 weeks, residue was preserved by ensiling and readily eaten by steers (Chapman et al., 1972). Orange peel residue at 80% moisture ensiled in airtight containers lost 25% dry matter through gas release and seepage during ensiling processes, contributing to increased costs on a dry matter basis (Ashbell and Donahaye, 1986; Ashbell and Weinberg, 1988). Mixing of mandarin residue with barley and carrots was reported to make nutritious and palatable silage (Onodera et al., 1996). However, handling and distribution problems for wet residue, press cake, and silage make these products economically unattractive compared with dried pulp or pellets.

10.2 MOLASSES

The pressing operation generates the press cake for dehydration and the press liquor for evaporation to molasses. Traditional high-temperature continuous evaporators at one time received the dilute press liquor and concentrated it to 72 °Brix, a concentration at low enough water activity to inhibit fermentation and allow storage at ambient temperature in iron tanks (Hendrickson and Kesterson, 1971). Very small quantities of traditional 72 °Brix molasses are now manufactured (30,000 mt; FCPA, 1997) since current industry practice involves use of waste evaporators to concentrate the press liquor to molasses in the 40–50 °Brix range. Use of 72 °Brix molasses is almost exclusively for dilution and fermentation to alcohol, although a small amount is used for formulation of sweetened cattle feed. Nutrient composition and properties listed indicate its suitability for these applications, with sugars being the primary nutrient (Table 10.4). The main use of waste heat evaporator molasses is for addition to the peel residue before drying, with some quantities sold as an alcohol fermentation substrate.

10.2.1 Processing

10.2.1.1 Press Liquor   The press liquor from the liming and pressing operation is generally 8–15 °Brix depending on the fruit source and whether plant wastewater is added. The soluble solids are composed of 60–75% sugars, mostly glucose, fructose, and sucrose with small amounts of pentoses (Table 10.5). The press liquor has a high biological oxygen demand (BOD) between

TABLE 10.4   Typical Composition of 72 °Brix Citrus Molasses[a]

Parameter	Mean	Parameter	Mean
pH	5–6	K (%)	1.1
°Brix	72	Ca (%)	0.8
Sucrose (%)	20.5	Na (%)	0.3
Reducing sugars (%)	23.5	Mg (%)	0.1
Protein (%)	4.1	Fe (%)	0.04
NFE (%)	62	P (%)	0.06
Fat (%)	0.2	Mn (%)	0.002
Fiber (%)	0.0	Cu (%)	0.003
Ash (%)	4.7	SiO_2 (%)	0.04
Pectin (%)	1.0	S (%)	0.2
Pentosans (%)	1.6	B (ppm)	6
Moisture (%)	28	Niacin (ppm)	35
Flavonoids (%)	3.0	Viscosity (cp)	2000–15,000

[a] Values adapted from Hendrickson and Kesterson (1971).

40,000 and 100,000 ppm because of the sugars and peel oil content and can be an environmental disposal problem if not concentrated to molasses (Kesterson and Braddock, 1976). Also, because of the high calcium content, scale is formed as insoluble calcium citrate and hesperidin precipitate in the tube nests, as the press liquor is concentrated in the evaporators.

The press liquor has some features that create handling problems. The surge tank feeding the evaporator may be >180,000 L (50,000 gal), and, since raw press liquor in the tank is contaminated with yeast and other microbes, fer-

TABLE 10.5   Composition of Citrus Press Liquor from Limed Residue[a]

Parameter	Mean	Range
pH	5.7	5.4–7.5
°Brix	10.1	6.1–12.6
Sucrose (%)	2.4	1.2–3.1
Reducing sugars (%)	4.2	2.8–5.8
Protein (%)	0.5	0.4–0.6
Pectin (%)	0.7	0.3–0.9
Pentosans (%)	0.3	0.2–0.4
Ash (%)	0.7	0.4–0.9
Citric acid (%)	0.2	0.2–0.3
Acetic acid (%)	0.1	0.1–0.8
Ethanol (% v/v)	0.2	0.1–0.5
Peel oil (%)	0.2	0.1–0.6

[a] Values adapted from Hendrickson and Kesterson (1971).

mentation may occur. If 40–50 °Brix molasses is made, it too may ferment since the water activity is high enough to allow fermentation. Fermentation of these materials uses the most fermentable sugars (glucose, fructose, and sucrose) first, creating alcohol and CO_2. In the evaporator, the alcohol is volatilized and may end up in the condensate water, while the CO_2 has to be dealt with in the noncondensable gas stream. The significant feature of this discussion is that dry solids yield is lost as a result of these sanitation practices. By calculation, alcohol fermentation efficiency is 50%, 10 °Brix press liquor has 0.1 g fermentable sugar/L (0.865 lb/gal). If 190,000 L (50,000 gal) press liquor is allowed to ferment to 0.5% alcohol (Table 10.5), 1800 kg (4000 lb) of solids will be lost and not sold as cattle feed or to the distiller. This is approximately 10% of the solids in the tank. For distillers, the problem is complicated by the fact that some of the most fermentable sugars have been lost, even though the molasses may be 50 °Brix. Solutions for the citrus processor are simple but costly; the press liquor may be pasteurized or chemicals added (e.g., bisulfite). The common recommendation is to keep the throughput as rapid as possible and occasionally clean and sanitize the system.

10.2.1.2 *Waste Heat Evaporation*

10.2.1.2 Waste Heat Evaporation Similar to juice evaporators, waste heat evaporators operate by feed forward flow of press liquor through the stages to the concentrate stage. The energy driving these evaporators is dryer exhaust gas steam created by evaporation of the water in the residue in the peel dryer. This concept was first reported as an energy saving feature for peel dryers (Snyder, 1967) and is now the accepted method for citrus processing plant feed mill operation. The saturated vapors from the dryer are sent to the evaporator at an initial temperature with the dew point of 82°C (180°F). These exhaust gases must be scrubbed to reduce particulate emissions before atmospheric discharge and to prevent scale formation on the outside of the tubes. Scrubbing is accomplished by recirculating first-effect condensate through nozzles located in the duct connecting the dryer exhaust to the evaporator and on top of all the first-effect bodies. Excess condensate flows out of the system. It is important that lime use be optimized to minimize scaling on the outside of the tubes from dust in the exhaust and from the press liquor on the inside of the tubes. Press liquor is commonly screened to reduce the fine particles, which contribute to fouling and increasing molasses viscosity.

The energy savings of waste heat evaporation result by shifting evaporation from the dryer to the evaporator. At start-up, press liquor is the °Brix of the limed peel residue/press liquor equilibrium, with press cake entering the dryer at 70–72% moisture and press liquor near 10 °Brix enters the evaporator. After this process reaches equilibrium, the press liquor (molasses) at 40–50 °Brix out of the evaporator is added back to the peel residue before pressing, which results in 20 °Brix press liquor now feeding the evaporator and 60–65% moisture press cake in the dryer (Fig. 10.1). At the lower moisture load, the dryer capacity increases. The fuel for this process is the energy to the dryer furnace.

10.2.2   Storage

As discussed above, waste heat molasses has limited storage life because of its propensity to ferment. This product, if filled into tank trucks has been known to ferment and foam during transportation to its destination. The traditional 72 °Brix molasses also has some stability problems related to foaming during storage. The low water activity of this product will not support growth of microbes; thus, it is usually stored at ambient temperature in large tanks. Occasionally, tanks of molasses begin to foam violently, creating steam and acrid vapors and considerable heat, with product overflowing and creating environmental problems. The exact cause of these occurrences has not been identified, although studies have postulated the chemical, sugar-amine browning reaction is responsible (Moyer, 1975). This reaction produces heat and CO_2 and seems to occur most frequently when fresh and old molasses are mixed, and has been noted for cane, beet, and citrus molasses.

10.2.3   Limonene Recovery

The press liquor and cake contain considerable peel oil (Table 10.5), which is volatilized during concentration in the waste heat evaporator and in the dryer exhaust. Similar to essence recovery in the juice evaporator, vapor from the first-effect tubes enriched in peel oil is condensed and flows to a tank at the bottom of the evaporator. This tank contains baffles and material for increasing surface area to break the peel oil condensate emulsion, allowing the oil to float to the top and be collected by decanting. The underflow from this decant tank is sent to waste treatment or to a spray field. The product oil is primarily composed of the chemical d-limonene, which is the name by which it is traded. An older synonym, stripper oil, derives from the steam stripping process used in conjunction with the traditional 72 °Brix molasses process. Inefficient recovery of peel oil and d-limonene allows escape into the waste heat evaporator/ dryer vent stack and contributes to release of volatile organic compounds (VOCs) into the atmosphere, a process that must be permitted by government regulation in Florida (Buff, 1997). Additional discussion of d-limonene is presented in Chapter 12.

REFERENCES

Ammerman, C. B., Hendrickson, R., Hall, G. M., Easley, J. F., and Loggins, P. E. (1965). The nutritive value of various fractions of citrus pulp and the effect of drying temperature on the nutritive value of citrus pulp, *Proc. FL State Hort. Soc.* **78**:307–311.

Arnold, P. T. D., Becker, R. B., and Neal, W. M. (1941). The feeding value and nutritive properties of citrus by-products. II. Dried grapefruit pulp for milk production, FL Agr. Exp. Sta. Bull. 354, University of Florida, Gainesville, FL.

Ashbell, G. and Donahaye, E. (1986). Laboratory trials on conservation of orange peel silage, *Agr. Wastes*. **15**:133–137.

Ashbell, G. and Weinberg, Z. G. (1988). Orange peels: The effect of blanching and calcium hydroxide addition on ensiling losses, *Biol. Wastes*. **23**:73–77.

Becker, R. B., Davis, G. K., Kirk, W. G., Arnold, P. T. D., and Hayman, W. P. (1954). Citrus pulp silage, FL Agr. Exp. Sta. Bull. 423, University of Florida, Gainesville, FL.

Braddock, R. J. and Cadwallader, K. R. (1992). Citrus by-products manufacture for food use, *Food Technol.* **46**(2):105–110.

Braddock, R. J. and Miller, W. M. (1978). Some moisture properties of dried citrus peel, *Proc. FL State Hort. Soc.* **91**:106–109.

Braddock, R. J., Kesterson, J. W., and Miller, W. M. (1977). Efficient processing of orange juice extractor residues into by-products, *Proc. Int. Soc. Citriculture* **3**:737–738.

Braverman, J. B. S. (1949). *Citrus Products; Chemical Composition and Chemical Technology*, Interscience, New York.

Buff, D. A. (1997). VOC emissions from citrus processing plants, *Trans. Citrus Eng. Conf.* **43**:65–79.

Chapman, H. L., Ammerman, C. B., Baker, F. S. Hentges, J. F., Hayes, B. W., and Cunha, T. J. (1972). Citrus feeds for beef cattle, FL Agr. Exp. Sta. Bull. 751, University of Florida, Gainesville, FL.

Cook, R. W. (1983). Multiple stage pressing, *Trans. Citrus Eng. Conf.* **29**:1–7.

Cook, R. W. (1987). Citrus peel dryer control system, *Trans. Citrus Eng. Conf.* **33**:53–56.

Covington, R. O. (1983). Steam dryer aimed at by-products, *Food Eng.* **55**(2):102–103.

Coxe, E. F. (1987). Cogeneration and the citrus industry, *Trans. Citrus Eng. Conf.* **33**:8–16.

FCPA (1997). Statistical Summary, 1996–1997 Season, FL Citrus Proc. Assn., Winter Haven, FL.

Heid, J. L. (1945). Drying citrus cannery wastes and disposing of effluents, *Food Ind.* **17**:1479–1483.

Hendrickson, R. and Kesterson, J. W. (1971). Citrus molasses. FL Agr. Exp. Sta. Tech. Bull. No. 677, University of Florida, Gainesville, FL.

Hillis, W. G., Ammerman, C. B., Loggins, P. E., and Hendrickson, R. (1968). Citrus pulp dehydrated with the aid of magnesium oxide and its subsequent use as a feedstuff. *Proc. FL State Hort. Soc.* **81**:297–301.

Johnston, R. B. (1995). Latest developments in pressing citrus peel, *Trans. Citrus Eng. Conf.* **41**:60–74.

Johnston, R. B. (1998). Pumped peel—five years later, *Trans. Citrus Eng. Conf.* **44**:79–90.

Kesterson, J. W. and Braddock, R. J. (1976). By-products and specialty products of Florida citrus, FL Agr. Exp. Sta. Tech. Bull. No. 784, University of Florida, Gainesville, FL.

Kesterson, J. W., Crandall, P. G., and Braddock, R. J. (1979). The heat of combustion of dried citrus pulp, *J. Food Process Eng.* **3**:1–5.

Leo, M. A. (1982). Energy conservation in citrus processing, *Food Technol.* **36**(5):231–233, 244.

Marques, D. S. (1995). Short time reaction system for improved citrus peel processing, *Trans. Citrus Eng. Conf.* **41**:92–105.

Mead, S. W. and Guilbert, H. R. (1926). The digestibility of certain fruit by-products as determined for ruminants. I. Dried orange pulp and raisin pulp, Agr. Exp. Sta. Bull. 409, University of California, Berkeley, CA.

Miller, W. M. and Braddock, R. J. (1982). Microwave drying of citrus peel, *Proc. Fla. State Hort. Soc.* **95**:204–207.

Moyer, C. E., Jr. (1975). Molasses stability, *Sugar J.* Jan. 20–23.

Neal, W. M., Becker, R. B., and Arnold, P. T. D. (1935). The feeding value and nutritive properties of citrus by-products. I. The digestible nutrients of dried grapefruit and orange cannery refuse and the feeding value of the grapefruit refuse for growing heifers, Fla. Agr. Exp. Sta. Bull. 275, University of Florida, Gainesville, FL.

Odio, C. E. (1993). Tank reaction system for citrus peel, *Trans. Citrus Eng. Conf.* **39**: 1–14.

Onodera, R., Kawamura, O., Inazawa, A., Izumi, T., Okuda, M., Katayama, H., and Yokoyama, M. (1996). Preparation of silages consisting of barley shochu distillery by-product and pulps of mandarin oranges and carrots, *Bull. Fac. Agric., Miyazaki Univ.* **43**(2):145–150.

Parish, M. E. and Higgins, D. P. (1989). Yeasts and molds isolated from spoiling citrus products and by-products, *J. Food Protect.* **52**(4):261–263.

Ross, I. J. and Kiker, C. F. (1967). Some physical properties of dried citrus pulp, *Trans. ASAE* **10**:483–485, 488.

Scott, J. J. (1926). Grapefruit refuse as a dairy feed, Fla. Agr. Exp. Sta. Ann. Rpt. 25R-26R, University of Florida, Gainesville, FL.

Sinclair, W. B. (1961). *The Orange: Its Biochemistry and Physiology*, University of California Press, Berkeley, CA.

Sinclair, W. B. (1972). "Grapefruit by-products." In *The Grapefruit*, University of California Press, Riverside, CA, Chap. 6, pp. 502–551.

Snyder, R. P. (1967). Waste heat recovery from feed mill dryers, *Trans. Citrus Eng. Conf.* **13**:28–43.

Solis, G. (1997). Private communication, Ag-Bag Intl., Ltd. El Paso, TX.

Sykora, G. (1997). Feed mill automation, *Trans. Citrus Eng. Conf.* **43**:1–14.

Timmons, D. E. (1950). Citrus canning in Florida: Early history and current statistics, AE Series 50-4, January, Fla. Agr. Exp. Sta., University of Florida, Gainesville, FL.

USDA (1962). *Chemistry and Technology of Citrus, Citrus Products, and By-products*, Agriculture Handbook No. 98, USDA, Washington, D.C.

Walker, S. S. and McDermott, F. A. (1917). The utilization of cull citrus fruits in Florida, Fla. Agr. Exp. Sta. Bull. 135, University of Florida, Gainesville, FL.

CHAPTER 11

ESSENTIAL OILS AND ESSENCES

Citrus peel contains oblate-shaped oil glands in the flavedo, which extend to different depths into the albedo of all citrus fruits. These cells contain chemicals discharged by metabolic processes of the fruit, which give each type of citrus its characteristic aroma and flavor. The prominent chemical classes present in these oils are terpenes, and the hydrocarbon, d-limonene, is the major constituent. The character of the flavor is mostly dependent on oxygenated terpene derivatives, aldehydes, ketones, esters, alcohols, and acids of the oils. The value of these oils for food and beverage flavorings, perfumery, and chemical uses has long been recognized. The large volume of published information and scientific literature about citrus essential oils are impractical to discuss in a single chapter. Therefore, the emphasis here will be related to the important technological issues of oil recovery, composition and properties, process variables, and products.

11.1 RECOVERY

The attentive student may find documentation of the early history of citrus oil recovery in many sources. Commercial quantities of lemon oil were made for centuries in the Motherland of essential oils (Sicily) by pressing peel in wooden screw presses or by hand squeezing oil-saturated sponges touched against scarified fruit. Classic essential oil reference volumes describe these processes and characterize the oils from citrus (Guenther, 1949). Early commercial sweet-orange oil recovery in the United States by the hand-sponge technique, vacuum distillation of aqueous extracts, and by pressing water-extracted peel has been

149

described (Hood and Russell, 1916). Later publications described citrus oil recovery, mechanical techniques, and oil properties (Kesterson and McDuff, 1948; Braverman, 1949; Kesterson et al., 1971).

11.1.1 Extraction

Current citrus industry recovery involves oil extraction machinery associated with the specific type of juice extractor in use. With a few exceptions, peel oil extraction is concurrent with the FMC, Brown, or Italian juice extractor systems (Chapter 4). Good descriptions of these oil extraction systems and information about the technology and machinery used has been reviewed (Huet, 1991). Universally, water is the solvent or carrier for the peel oil as it is extracted from the oil glands and enters the recovery process. Since the oil components are mostly insoluble in water, recovery involves handling oil–water emulsions and homogeneous suspensions (Matthews and Braddock, 1987). For efficiency, modern processes recycle the water within the oil recovery systems.

11.1.1.1 FMC Process   The FMC oil recovery procedure is unique because the juice extractor contains the components necessary to rupture the oil glands and extract the oil into the emulsion. This process minimizes the space and energy for high yields of oil from the peel, at the same time extracting the juice (FMC, 1998). Oil extraction occurs in sequence after fruit is placed on the lower extractor cups and cutter tubes, the upper cups descend, and plugs are cut in the fruit from the bottom cups. The upper cups descend, forcing the juice and inner fruit contents down through the bottom tube for juice recovery. At the same time, the peel is shredded, being forced through openings in the upper cups, which ruptures the oil glands. The upper cups contain spray rings, which apply pressurized water to the peel during and after the shredding step, emulsifying the oil as it is released. This emulsion and small particles of peel and soluble and insoluble solids flow from the extractor, are collected, and sent to a finisher for initial separation of the larger particles. The finished emulsion is sent to a centrifugation process for concentration and recovery of the final cold-pressed oil. The water discharge from the centrifuges is filtered and recycled to the process.

Process variables may affect oil yield and quality. Generally, fruit size, condition, and quality are of major importance, where firm, mature fruit that fits properly into the extractor cups produces highest oil yields. Enough water also must be sprayed onto the shredded peel to adequately carry the emulsion, but not so much that emulsion oil concentration will require extensive centrifugation to recover the oil. The amount of water used in the extraction process generally does not affect the oil quality (aldehydes) as much as the fruit maturity (Steger, 1981). Peel sugars dissolve in the water during recycling and may concentrate to the range of 4–10 °Brix, before bleed-off to the feed mill waste stream. Good sanitation is a requirement for recycle systems since the incoming fruit provides inoculum for microbial growth on the sugars, which in turn can result in fermented flavors in the cold-pressed oil. Water use for orange peel oil recovery may average 10–11 L/box (3 gal/box) at >50% oil recovery extraction efficiencies of the amount in the fruit (Steger, 1979). Some amount of water is absorbed by contact with the peel in this process and must be evaporated in the feed mill.

11.1.1.2 Brown Oil Extractor (BOE) and Shaver The Brown juice extraction process (Chapter 4) maintains two oil extraction systems, the peel shaver and the BOE. The shaver has largely been replaced by the BOE, but is still available for certain by-product applications and for tangerine oil extraction, after juice recovery with Brown 700 or 1100 extractors (Chapter 4). This machine operates by shaving the flavedo from the albedo of the peel from the extractors, spraying water onto the flavedo as it is pressed between knurled rolls, releasing the oil into the water to obtain the emulsion. The knives may be set to different thickness for degree of flavedo shaving into the albedo layer. Separate by-product streams of flavedo and albedo may be obtained in addition to the oil emulsion. Optimal shaver setup involves placing a machine beneath each juice extractor.

The BOE operation requires the machine to be installed in the fruit stream after washing and before the juice extractors. This machine extracts the oil by passing fruit over a series of rotating toothed rollers in a pool of water (Bushman, 1972). The oil glands are cut, releasing the oil into the water, and the fruit passes on to the juice extractor. Although this is a scarification process, very close inspection of the fruit surface is required to detect the cuts. An improvement, allowing movement of blocks of fruit at a time, is claimed to increase oil yield (McKinney, 1981).

Water is typically recycled, reducing water usage, to a °Brix where bleed to the feed mill occurs. Because only the flavedo is contacted by water, the peel does not absorb much water and the emulsion water °Brix increases slowly during recycling, resulting in low viscosity of the stream feeding the centrifuges. In a recycle system, the BOE process will require 1–4 L/box (0.3–1 gal/box) makeup water added as a fruit rinse at the end of the process (Waters, 1993). Low amounts of insoluble solids in the emulsion stream increase oil recovery efficiencies in the centrifugation steps. The BOE may be set up for efficient oil extraction from the different fruit varieties, reducing peel oil concentration in the juice during extraction, provided the fruit surface is adequately rinsed after oil extraction. A necessary requirement for efficient operation of this machine is to prevent fruit rate overload; otherwise, the surface will not contact the rollers and release the oil. A complete description of this process and its operation to recover >50% of the oil in Valencia oranges has been published (Waters, 1993).

11.1.1.3 Italian Machines Some peel oil is recovered from whole fruit by a scarification process, whereby whole fruit enter a cylindrical chamber containing spinning sharp perforated rollers. As the fruit pass down through the machine, the flavedo is scarified, releasing the oil into a water spray inside the chamber. The oil–water emulsion collects in the bottom and is sent to the finishing/centrifugation process, while the fruit proceeds on to the juice extractors (Chapter 4). An alternate process extracts oil from peels of half fruit (or whole fruit, including the juice) by mixing with water in a screw press. The press liquid is the oil emulsion stream, and the press cake may be mixed with more water and pressed again to increase yield. This emulsion may be centrifuged for cold-pressed oil recovery or sometimes, distilled to recover the oil. These processes produce oils high in pigment, wax, and residue, and they have specific demands by the flavor industry, different from oils extracted by the Brown or FMC processes.

11.1.2 Centrifugation

Following a finishing step to remove larger insoluble pieces of peel, cold-pressed oil recovery from emulsions of the extraction processes involves a two-stage centrifugation process. Operation of the entire process may be automated, and efficiency of the extraction/centrifugation process determines the oil yield; however, the primary efficiency limiting step is the extraction from the peel. In some cases (tangerine oil), the finishing may involve heating the peel/oil emulsion in the finisher. This helps extract oil from the peel particles and increases centrifuge efficiencies due to lower viscosity of the hot emulsion. The process is defined in the flow diagram of Figure 11.1, showing the initial oil concentration in the emulsion, 0.5–3.0%, is concentrated to 50–70% by the separator. The concentrated emulsion is fed to the polisher and separated into pure oil and a waste sludge. The first-stage separator, known as the desludger, may operate between 7000 and 10,000 times gravity at 100–400 L/min (25–100 gal/min), while the much smaller polisher operates at similar g forces at lower feed rates of 5–10 L/min (1.3–3 gal/min) (Matthews and Braddock, 1987).

The maximum oil recovery efficiency depends on each operation since the overall efficiency is a multiple of the efficiency of each process. Excellent discussions of the variables relating yield, extraction, and centrifugation of peel oil emulsions to process efficiency has been published for the BOE and shaver (Waters, 1993) and for the FMC process (Ferguson, 1980). In general, the simplest calculation of efficiency works for each of these processes, as follows:

Total process efficiency (eff) = (extraction eff) $\times$ (separator eff) $\times$ (polisher eff)

Extraction eff(%) = 100(MO)/WF

Separator eff(%) = 100(F − AQ)/F

Polisher eff(%) = 100(CE − S)/[CE − ($S \times$ CE/100)]

where MO is the mass oil in extracted dilute emulsion, WF is the mass oil in whole fruit, F is the percent oil in feed stream, AQ is the percent oil in aqueous

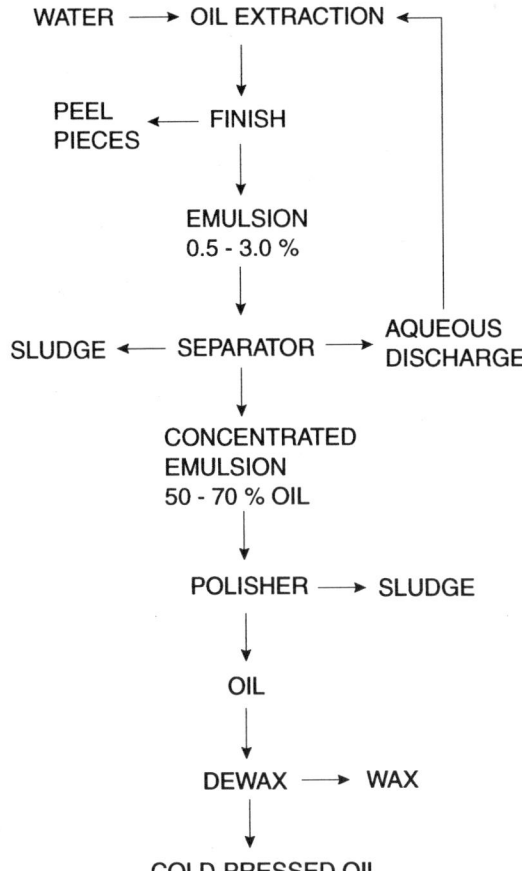

Figure 11.1 Flow diagram of a typical citrus cold-pressed oil recovery and water recycling process.

discharge, CE is the percent oil in concentrated emulsion from separator, and S is the percent oil in sludge.

The extraction efficiency is not simple to calculate without some guessing for the various extraction processes. For all methods, fruit flow must be estimated, but other than the BOE, oil in the various component fractions must be determined, as well as an estimate of their mass flow rates. Under ideal conditions, it should be possible to extract 60–75% of the oil from the fruit and to actually recover 95% of the amount extracted. Picking, hauling, loading, and unloading fruit causes loss of oil cell turgor and subsequently lowers oil recover compared with fruit fresh from the tree. Also, economic value may help determine process efficiency. One would tend to try harder to recover lemon oil than orange oil. However, these calculations help monitor equipment operation performance, even with many variables affecting each process. Full understanding is required for optimum benefit and the bottom line is the amount of total oil in the fruit actually recovered.

11.1.3   Total Oil in Fruit

11.1.3.1   Quantity   Amounts of peel oil in the various citrus varieties were determined by early extaction procedures for lemon and orange oils. Hood and Russell (1916) reported that 3.5 kg oil/mt fruit (7 lb oil/ton) could be obtained from oranges by a scarification/pressing process. Also, ranges of 4–10 kg/mt (8–20 lb oil/ton) for California lemons and 2–4.5 kg/mt (4–9 lb/ton) Valencia oranges by both hand pressing–sponge and machine methods were reported (Poore, 1932). Theoretical values of the total oil available for recovery in the various commercial Florida citrus varieties have also been determined (Kesterson and Braddock, 1975). These values, presented in Table 11.1, support the historic and modern commercial 60–75% yields of oil in the fruit obtained by recent manufacturing processes. Examination of the results clearly shows that tangerines, Valencia oranges, and lemons have more peel oil than grapefruits, early–midseason oranges, Temples, and Persian limes.

11.1.3.2   Horticultural Variables   The importance of horticultural practices to crop yields and fruit and juice quality is recognized. However, there have not been many studies in relation to peel oil content and quality. Considering the surface of an individual fruit, more oil was found in the bottom (stylar) half and south-facing part of California Valencia oranges (Bartholomew and Sinclair, 1946). A comparison of peel oil content from over 30 Valencia budwood selections budded to one type of rootstock found values ranged from

TABLE 11.1   Peel Oil Content of the Major Citrus Cultivars

Cultivar	Samples[b]	Lb Oil/Ton Fruit[a]			
		Min.	Max.	Mean	SD
Hamlin	64	7.0	8.2	7.8	1.5
Pineapple	100	7.1	14.0	9.7	1.0
Valencia	255	10.4	16.3	13.5	1.8
Dancy tangerine	45	13.5	17.4	15.5	1.8
Temple	47	6.8	9.1	7.9	0.7
Orlando tangelo	36	9.7	12.7	11.3	0.7
Duncan grapefruit	83	4.8	6.9	5.6	0.6
Marsh grapefruit	87	5.5	7.3	6.2	0.5
Ruby grapefruit	82	5.1	7.8	6.5	0.5
Bearss lemon	270	11.9	19.2	15.1	2.4
Persian lime	121	7.3	9.2	8.1	1.2

[a]Multiply by 0.5 to convert lb/ton to kg/mt.
[b]Total samples for four seasons (Kesterson and Braddock, 1975).

5.5 to 8 kg/mt (11 to 16 lb/ton fruit). This study also reported that the same budwood scion on 19 rootstocks showed oil yield was also affected by the rootstock, although not to the degree as the budwood selection (Hendrickson et al., 1970). Budwood selected from clones with high fruit yields showed oil content increases of 2 kg/mt (4 lb/ton) of lemons and 2.3 kg/mt (4.6 lb/ton) of Valencia oranges, compared with the standard varieties (Kesterson and Braddock, 1977). Fertilization studies on lemons and pineapple oranges have also demonstrated that increased nitrogen increased oil yield, while increased application of potassium suppressed oil yield (Kesterson and Braddock, 1977; Kesterson et al., 1977). Statistically significant differences in shape, rind thickness, juice yield, acid content, and color were determined from lemons from 11 countries, illustrating the diversity of this fruit (McDonald and Hillebrand, 1980).

The aldehyde content of orange oil is a primary indicator of flavor strength, and higher is better. A study of the effect of fruit maturity on the chemical and physical properties of Valencia oil reported that oil from immature (green) fruit had 1.3% aldehydes and lacked fresh character, while oil from mature fruit had 1.8% aldehydes and good aroma (Kesterson and Hendrickson, 1962). Similar studies for grapefruits have reported that aldehydes and nootkatone contents increased until the fruit was past peak maturity (Kesterson et al., 1965). Citrus fruit at the peak of maturity and firmness should have the highest oil yield, at maximum turgidity of the oil glands. It is after this time the flavor and aldehyde contents of the oils are highest.

11.1.4 Juice Oil

Considerable peel oil may find its way to the juice during juice extraction. Since most consumers prefer juice to have less than 0.025–0.030% oil, excess oil is removed from NFC juices (Chapter 5) before packaging. Depending on fruit type and extractor setting, juice peel oil content after the extractors may be as high as 0.06% by volume in situations where maximum juice recovery is the objective. The NFC juice oil reduction to 0.020% may be done by extractor setting to less pressure conditions, or by vacuum/steam stripping part of the volatiles. Economically, lower extractor and finisher settings stipulate lower juice yields, and vacuum deoiling may lose desirable volatile flavors.

An older technology, which has seen revival since the advent of aseptic NFC juice, allows removal of excess juice oil by centrifugation. The many ways centrifuges can be used for pulp reduction, debittering, viscosity control, and the like in citrus plants have been documented (Distelkamp, 1962). A process scheme to recover juice oil by centrifugation claimed to produce a fresh-flavored juice and oil with much better top-notes than oil recovered by vacuum stripping of single-strength juice (Thrush, 1964). The centrifugation process to recover juice oil involves separation of a concentrated emulsion from the juice, where as much as 50–65%, v/v, oil reduction occurs. This concentrated emulsion may be clarified in a small polisher, similar to cold-pressed oil recovery. The oil recovered may amount to as much as 65% of the volume of oil reduction of the juice stream. The properties of juice oil are very similar to essence oil from juice evaporators, where esters and top-notes predominate over aldehydes. Juice oil from oranges has been described to have less aldehydes and more esters than cold-pressed oil (Kesterson and Braddock, 1976).

11.1.5 Wax

The nonvolatile peel cuticular wax coating, pigments, lipids, flavonoids, and some suspended water and other matter dissolved in the oil during extraction is termed wax. The wax is perceived as a defect in cold-pressed oil, which has been chilled, as it crystallizes and the flocculant has an unsightly appearance. The composition of this wax has been described for lemons and oranges (Sinclair, 1984) and for grapefruits (Markley et al., 1937). The cuticular waxes are primarily long-chain hydrocarbons and alcohols of carbon chains between C_{20} and C_{26}. A thorough study of the wax constituents of grapefruit peel has identified the triterpenoid, friedelin, as a major constituent, including the alkanes, alcohols, sterols, and squalene (Nordby and McDonald, 1994). Grapefruit trees conditioned in 15°C chambers against chilling injury contained higher quantities of squalene in the wax, evidence that this compound may protect fruit from chilling injury (Nordby and McDonald, 1990).

The cold-pressed oil recovery process involves winterization to reduce the wax content. Besides being unsightly in the oil, which may be added to clear beverages, wax can cause viscosity problems during vacuum distillation (folding) to remove terpenes from the oil. The conditions for winterization have not been well defined; however, the wax crystallizes from oil held at temperatures below 0°C for long periods (Hendrix et al., 1992). The lower the temperature, the shorter the time needed for dewaxing. Cold-pressed oil may be typically chilled to −25 to −40°C from 2 days to a week and then filtered through diatomaceous earth to remove the wax. Higher temperature (−15°C) from 20 days to several months may be used without filter aid, where the oil is decanted. Grapefruit oil is usually winterized by holding for 3 weeks at −24°C, then recovered by filtration.

Oil in bulk quantities is held in slender, conical bottom, stainless steel tanks to facilitate decanting the clear oil above the precipitated wax (Matthews and Braddock, 1987). The chilling process and cold storage of the oil require refrigeration energy. To cool cold-pressed oil from ambient temperature of the centrifuge process across a temperature differential of 20–60°C would require 36–110 kJ/kg oil, or about 0.7–2.1 kWh, based on the specific heat of d-limonene (1.834 kJ/kg K) (Braddock and Miller, 1982).

11.2 COMPOSITION

There have been many studies of the composition of volatile constituents in the various citrus oils. The interested reader may examine the studies listed

**TABLE 11.2   Approximate d-Limonene
Concentration of Various Citrus Essential Oils**

Product	Limonene (%, v/v)
Citrus terpenes, d-limonene[a]	>95
Orange, tangerine, tangelo[a]	95
Grapefruit[a]	93–95
Essence oil[a]	95
Lemon[a]	75–80
Lime, Mexican, Persian[a]	50–55
5-Fold orange, grapefruit[b]	90
10-Fold orange[b]	80–85
25-Fold orange[b]	60–65
36-Fold orange essence oil[a]	1–2

[a]Braddock et al. (1986).
[b]Vora et al. (1983).

here for reference and pursue the literature for in-depth information. A complete database of the individual composition and concentrations of the classes (hydrocarbons, carbonyls, etc.) of volatile compounds for most of the citrus oils and juices has been compiled (Maarse and Visscher, 1989). This useful work is now probably the standard for identity and quantity of the hundreds of volatile flavor components found in the individual citrus fruit. This database is complemented by a thorough review of the individual chemicals and chemical classes considered important to flavors from oranges, tangerines, mandarin fruits, grapefruits, lemons, limes, and kumquats (Shaw, 1991). Citrus oil composition of interest for commercial purposes involves physical and chemical properties as well as typical aroma and sensory evaluation. While oils from the various citrus cultivars have distinct aroma and flavor specific to their fruit source, these oils are surprising in their chemical similarity. The oils from the various oranges, grapefruits, tangerines, and hybrids have d-limonene contents greater than 90%, while lemon and lime oils may have less than 75% d-limonene (Table 11.2). All citrus oils contain over 90% monoterpenes, considering both d-limonene and oxygenated terpenes.

11.2.1   Orange Oil

The chemical and physical property standards of identity for cold-pressed orange oil are defined in individual monographs of the various citrus oils (Food Chemicals Codex, 1996). These standards (e.g., cold-pressed orange oil) establish analytical methods and ranges for specific gravity (0.842–0.846), refractive index (1.472–1.474), optical rotation (+94° to +99°), aldehyde content (1.2–2.5%), and heavy metals. Specific gravity, optical rotation, and refractive index of the oil are properties, measured historically, with the goal of adulteration detection. These values still find use in evaluation of oil properties for flavor emulsions and formulations related to specific oil characteristics. The flavor strength of the oil, estimated by measurement of the aldehyde content by quantitative wet chemistry, is also useful for economic reasons in the oil trade and for product formulation. Aldehyde contents of early and midseason orange oils are generally in the lower part of the Codex standard range, while Valencia oils are higher.

The aroma and flavor significance of the volatile compounds of the various orange oils as used in juices, beverages, and foods is undeniable. However, consider the challenge to flavor formulators using quantitative data of the individual components listed in the database (Maarse and Visscher, 1989). For example, 14 references in the orange peel oil database listed concentrations from 360 to 4000 ppm for the hydrocarbon, valencene, an important flavor component. Even more significant is the value range of 1700–22,000 ppm for decanal, a major aldehyde contributing to the total aldehyde content (by the Codex analysis). The Codex range of 1.2–2.5% total aldehydes would dictate that values below 10,000 ppm in the literature might be erroneous.

The importance of citrus oil oxygenated components have resulted in studies to better define quantitative composition of the oils. The major aldehydes of the oils are the homologous series of aliphatic C_2 units and terpene aldehydes, made by biochemical synthesis in the fruit. The concentrations of orange oil aldehydes were determined by chemical and crystallization methods as 31% octanal, 27% decanal, 6% dodecanal, and 7.5% citral (Naves, 1947). Decanal and octanal are the primary aldehydes of the citrus oils with the exception of lemon and lime oils, which contain citral (neral + geranial). Aldehyde content of commercial Florida oils, determined by quantitatively preparing the 2,4-DNPH derivatives is reported in Table 11.3 (Braddock and Kesterson, 1976).

TABLE 11.3   Concentration of Major Florida Citrus Oil Aldehydes as Dinitrophenylhydrazone (DNPH) Derivatives[a]

Oil	C_6	C_8	C_{10}	C_{12}	C_{14}	Neral	Geranial
	% (wt/wt) of Total Aldehydes						
Hamlin	1.2	29.2	22.7	15.1	6.1	1.1	9.3
Pineapple	0.5	28.2	18.0	9.9	6.5	6.9	10.1
Valencia	1.0	27.3	30.7	9.4	5.3	4.1	4.4
Valencia (Calif.)	Trace	20.0	37.2	9.8	4.0	5.8	7.4
Temple	Trace	32.2	17.9	10.7	5.4	3.6	4.5
Dancy tangerine	Trace	23.1	24.0	13.5	4.8	5.0	5.7
Orlando tangelo	Trace	26.0	23.6	12.2	10.1	9.0	6.0
Duncan grapefruit	Trace	29.9	17.8	12.7	6.9	2.1	7.1
Essence oil (Val.)	0.5	13.8	26.7	5.5	1.0	6.0	4.2

[a]Modified from Braddock and Kesterson (1976).

11.2.2 Tangerine (Mandarin) Oils

Since tangerine juice is less common than orange juice, fewer fruits are grown and processed, a requirement for recovering oil. Cold-pressed tangerine oil has the unique aroma of the mandarin fruits, different from orange, yet the composition of the major constituents is similar. The source of this aroma is in the minor constituents and compositional ratio of the major components, which have been reviewed (Lawrence, 1992). In different parts of the world, different mandarin cultivars are grown and each has a slightly distinct oil aroma. Of the oxygenated oil components responsible for this unique aroma, methyl-*N*-methylanthranilate, thymol and sesquiterpenes, α-sinensal, and β-sinensal are very important in the juice (Shaw, 1991, 1996) and in the oil (Moshonas and Shaw, 1974). Tangerine oils are also valued for their dark, reddish pigmentation, which is useful for blending natural color into certain beverage formulations.

11.2.3 Grapefruit Oil

Grapefruit flavor and aroma is quite unique and distinguishable from the other citrus varieties by being harsher and less easily affected by processing and storage (Shaw, 1991). Collected directly from the polisher, cold-pressed grapefruit oil lacks typical grapefruit aroma and, by odor, is difficult to distinguish from orange oil. Cold-pressed grapefruit oil flavor benefits from storage, or aging, to develop its distinctive grapefruit character. Also, except for color, there is little difference between white and red grapefruit oils. The oil is quite stable during storage for 6–12 months at 18°C, during which time the linalool content decreases and the full-bodied grapefruit aroma develops (Kesterson and Hendrickson, 1967). The stability during storage may be influenced by the high level of α-tocopherol (>250 ppm) in the cold-pressed oil, which is twice the amount in orange oil (Waters et al., 1976).

The grapefruit-like character of the oil and juice once was thought to be largely influenced by the compound (+)-nootkatone (Fig. 11.2) (Shaw, 1991). Many studies of the volatile juice constituents have established the difficulty of quantifying the highly volatile compounds (Cadwallader and Xu, 1994). However, studies by commercial researchers have defined the major character-impact components of grapefruit juice (and presumably, oil) to be terpene thiols. Identification of the compound, 1-*p*-menthene-8-thiol (Fig. 11.2) has resulted in the ability to synthetically reproduce the flavor of fresh grapefruit (Demole et al., 1982). Discovery that 1-*p*-menthene-8-thiol has an aroma threshold below 1 ppb makes this compound one of the most potent known natural aromas. A number of these thiols have been identified and synthesized by acid catalysis of terpenes with hydrogen sulfide (Janes et al., 1993). These compounds, as well as some sesquiterpenes (Demole and Enggist, 1983), similar to (+)-nootkatone, practically complete the definition of fresh grapefruit flavor. Mixtures of 1-*p*-menthene-8-thiol and α-terpineol added to β-

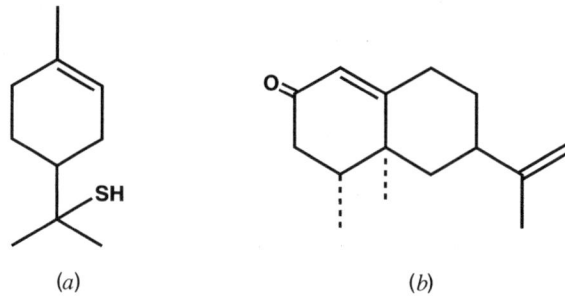

Figure 11.2 Components important to the flavor of fresh grapefruit juice. (*a*) 1-*p*-menthene-8-thiol and (*b*) (+)-nootkatone.

phenylethanol have been shown to have different aroma characteristics at different concentrations. At 1–5 ppm, the mixture has a rose petal aroma, and at 10^{-5} ppm the material has a guava, grapefruit-like tropical note (Mookherjee et al., 1985).

11.2.4 Lemon and Lime Oils

Lemon and lime trees are more temperate than the other citrus varieties, which limits their distribution to the warmer world citrus zones. The trees have more thorns than oranges, which makes picking the fruit difficult. Also, the juice, because of its acidity, is not as marketable as orange juice, all factors contributing to fewer plantings of these fruits. Since the oils are very prized for their flavor and aroma, scarcity of fruit and low demand for processed juice results in higher value of these oils compared to the other citrus oils (Table 11.4).

The aldehyde content of lemon oil is considerably higher than orange, ranging from 2.0 to 5.5% (as citral) (Food Chemicals Codex, 1996), depending on the type of lemon and the growing locale. Lime oil aldehyde content is even higher, with differences noted between cold-pressed and distilled West Indian (Mexican) and Persian (Tahitian) limes (Table 11.5). The d-limonene content of these oils is also significantly less than that of the other citrus oils. Oil from a lemon cross, the Meyer lemon, has an unusual aroma related to its composition of approximately 5% thymol (Moshonas et al., 1972); however, this oil has a low aldehyde content. Cold-pressed lemon and lime oils also contain significant amounts of coumarin-type compounds in the high-boiling fraction. Some of these coumarins and psoralens may cause photosensitive skin reactions in people who come in contact with the oil and go into the sunlight. Based on concentration of coumarins, particularly bergapten, in the peel, Persian lime oil may be potentially more phototoxic than West Indian lime oil (Nigg et al., 1993).

The lime oil of commerce is distilled West Indian oil, which is manufactured by distillation of a water extract of peel or whole, chopped, pressed fruit.

TABLE 11.4 Approximate Value of 1 lb of
Various Citrus Oils of Commerce[a]

Oil	$ Value
d-Limonene	0.45
Orange, Brazil Pera	0.65
Orange, FL midseason	0.70
Orange, FL Valencia	0.77
Orange, CA distilled	1.25
Orange, bitter	19.00
Grapefruit, FL	12.00
Tangerine, FL	14.00
Mandarin, Italian	35.00
Lemon, CA	9.50
Lemon, Argentina	12.50
Lemon, Italian	15.00
Lime, Mexican distilled	11.00

[a]Adapted from Chemical Market Reporter (1998).

Because of the small fruit size, the juice may not be recovered. However, juice recovered from the presses can be manufactured into a clarified juice product. The process involves clarifying by preservation with bisulfite, allowing cloud precipitation by pectinesterase, and traditional filtration with diatomaceous earth or by ultrafiltration. The clarified lime juice product is used in alcoholic drinks. Steam distillation of the oil occurs in a still with refluxing for 8–10 hr at pH of 2–2.5, with the oil recovered by decanting from the condensate. The refluxing is necessary to obtain the acid-catalyzed conversion of citral to *p*-cymene (and other changes) necessary to develop the unique flavor of the distilled lime oil. This conversion is reflected by lower aldehyde contents of the distilled oils presented in Table 11.4. The kinetics and compositional chemistry of these reactions has been thoroughly discussed in an excellent review (Clark and Chamblee, 1992). Even under rather mild temperature (37°C) stor-

TABLE 11.5 Properties of Cold-Pressed and Distilled Persian and West Indian Lime Oils

Property	Cold-Pressed		Distilled	
	Persian	W. Ind.	Persian	W. Ind.
Aldehydes (%)[a]	5.0	6.5	2.4	1.5
Specific gravity (20°C)	0.874	0.882	0.857	0.868
Refractive index (20°C)	1.483	1.483	1.475	1.477
Optical rotation (20°C)	+43.8	+37.5	+48.9	+46.0

[a]Expressed as citral (Kesterson et al., 1971).

age for 30 days, emulsions of lemon oil and citric acid solutions formed a number of intense off-flavor components at the expense of citral and citronellal (Schieberle and Grosch, 1988).

11.3   DISTILLED OILS

Citrus oils that are not cold-pressed and separated from emulsions by centrifugation because of their volatility may be recovered by distillation. Besides distilled lime oil (above), these oils include the volatile water and oil-phase essence from juice evaporators, d-limonene stripped during concentration of molasses, distilled oil emulsions, and the concentrated (folded) products obtained by distillation of cold-pressed oils.

11.3.1   Essence and Essence Oil

Volatile components in concentrated juice would be lost during evaporation without the vacuum stripping, condensation, rectifying, and concentration process referred to as essence recovery. Citrus fruit aroma is a complex mixture of highly volatile (acetaldehyde), moderately volatile (terpenes), and slightly volatile (sesquiterpenes) compounds. Certain constituents have intense flavor at very low concentrations (limonene thiols) or very little flavor at high concentration (ethanol). Some are soluble and desirable in the juice; others are not. Also, many of the compounds are sensitive to chemical, oxidative, thermal, or biochemical degradation, before, during, or after processing. With this scenario, the difficult objective of recovery processes is to obtain an aroma most characteristic of fresh juice.

As heated juice is flashed into the first (or other) effect of the evaporator (Chapter 5), the volatile aroma components and water are vaporized. The water becomes the steam for driving the evaporation by condensation in each successive effect. The volatile aromas in the juice vapor also pass under vacuum to the outside of the tubes of successive effects, where they are recovered. One of the early successful designs of citrus essence recovery involved trapping and condensing these entrained vapors (Brent et al., 1966). Essence recovery may also be achieved from raw fruit, juices, and vegetable products independent of juice recovery by a vapor–liquid spinning cone contacting process (ConeTech, Inc., Salinas, CA).

Most commonly, essence vapors are recovered from the first and second flash to the second and third effects of feed-forward evaporators, where concentrations are highest and temperature exposure is minimized. Inside the shell of the evaporator, volatile-containing vapors from the previous effect concentrate at some location, usually near the upper third, in the tube nests. At this point of maximum concentration, condensation and refluxing of water and the vapors may occur. It is at this point of a properly designed evaporator that volatiles are sent to the still (essence recovery unit) mounted near the top of the evaporator. There are many evaporator designs and ways to recover the volatiles; however, most are recovered in the second or third effects of the evaporator during removal of the first 25–30% of the juice concentration process. A complete description of the juice evaporation and essence recovery process has been published for the common citrus juice TASTE evaporators (Redd et al., 1996).

When a liquid containing multicomponents is heated to its boiling point, the vapor composition is normally different from that of the liquid. The compositional difference of the two phases at equilibrium forms the basis for separation and concentration. Since many of the aromas of juice are more volatile than water, a fractional distillation process is used to rectify and concentrate them and remove water. The vapor streams [Fig. 11.3, (1)] from the noncondensable gases are collected from the second and third effects of the evaporator and sent through a valve system to the essence recovery unit. These volatiles enter the stripping column (2) of the still below the overhead condenser (3), which aids in water removal from the volatiles. This water exits through the bottom dump. The enriched volatiles and some noncondensable gases pass to a chiller (4) and into a scrubber (5). Condensed volatiles from the scrubber drain to the bottom for collection and the remaining vapor passes to a cold condenser (6) for condensate collection in a decant tank (7) at the bottom of the evaporator. The condensed decant tank product contains aqueous and oil (terpene) phases that are typically separated by decanting the oil and pumping the aqueous phase to a tank in a refrigerated room (8). Sometimes the oil-phase

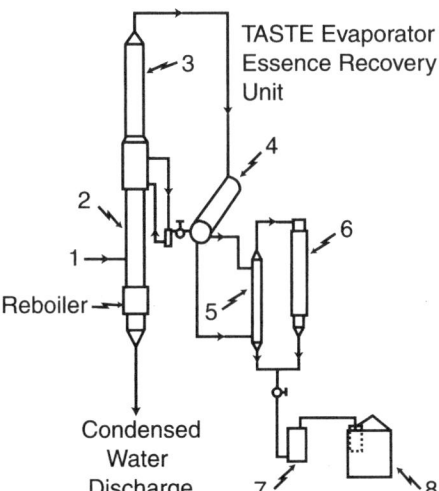

Figure 11.3 Generalized flow diagram of a TASTE evaporator citrus juice essence recovery system (Redd et al., 1996). Components include (1) vapor stream from evaporator effects, (2) stripping column, (3) overhead condenser, (4) chiller, (5) scrubber, (6) cold condenser, (7) decant tank, and (8) refrigerated tank.

essence (essence oil) is recovered by centrifuge from the cold-room product, for producing low oil, aqueous essence.

The process results in a 200- to 500-fold concentration of the volatiles, now called essence. Fold is the term used to measure the strength of the essence, for example, 1 gal of essence recovered from 200 gal of juice is defined as 200-fold. A rule-of-thumb mass balance for essence recovery yields approximately 100 lb aqueous essence (13% ethanol) and 7 lb essence oil from 1000 boxes of Valencia oranges (Johnson and Vora, 1983). The main aqueous-phase constituents of orange essence are water, >65%, ethanol, 12–25%, and methanol, 0.5–2%, depending on the condition of the juice and operation of the essence unit. The natural aqueous juice aroma constituents are present in the parts per million range, with acetaldehyde the major component. The essence oil is mostly d-limonene, 94%, valencene, >1%, myrcene, 1%, pinene, 0.5%, octanal, 0.5%, decanal, 0.6%, and so forth (Johnson and Vora, 1983). Ethanol in the essence is a major component of biochemical metabolism of the fruit and may approach 0.1–0.5% concentration in the juice from very overmature fruit. Juice methanol is a reaction product of pectinesterase activity, producing pectic acid and methanol from pectin, which also increases with fruit maturity (Chapter 5).

Essence oils and aqueous aromas have value for flavoring juices, concentrates, and beverages. Because aqueous orange essence contains >12% ethanol, but low amounts of natural aroma compounds, this product has less value than essence oil and is sometimes discarded. Acetaldehyde in the essence has value, as it helps impart fresh juice top-notes to fruit flavorings. It is possible to increase the acetaldehyde concentration by natural bioconversion from ethanol in the essence (see Chapter 14). It has been shown that acetaldehyde content of essence could be increased significantly by biomodification through use of the yeast *Pichia pastoris* (Goodrich et al., 1998). This yeast contains an enzyme complex, alcohol oxidase, that has the ability to oxidize ethanol to acetaldehyde, without requiring the expensive regenerative cofactors of alcohol dehydrogenase systems (Raymond, 1984).

Another method to concentrate aqueous essence components involves the use of reverse osmosis, with membranes capable of retaining small, highly volatile molecules such as acetaldehyde (Braddock et al., 1991). Retention of ethanol and acetaldehyde depended on the process temperature and concentration achieved but was greater than 50% at essence ethanol concentrations <10%. Larger, less volatile molecules (ethyl butyrate, hexanal, linalool, and limonene) were separated with >90% efficiency.

11.3.2 Folded Oils

11.3.2.1 *Distillation and Alcohol Washing*   Because pure d-limonene has little or no flavor character and may readily oxidize or develop undesirable "terpeney" off-flavors, citrus oils may be concentrated to remove a portion of the limonene. These concentrated oils have improved stability and solubility in

beverages (Redd et al., 1996). The process is called folding and is measured by the ratio of the reduction in volume of the single-strength oil. For example, 5-fold oil (20 mL) will result from single-strength oil (100 mL) by removal of mostly d-limonene. Folding is commonly performed by vacuum distillation or sometimes by alcohol washing. In the example of distillation, the highest vacuum, lowest temperature, shortest time to achieve the concentration desired is necessary to avoid flavor degradation of the oil. Process conditions for manufacture of folded citrus oils by fractional distillation and alcohol washing have long been established (Bennett, 1934; Guenther, 1949). Since d-limonene is insoluble in 60–70% ethanol, orange oil may be folded by liquid–liquid extraction with alcohol. This process is achieved by slowly stirring the oil–alcohol mixture, partitioning oxygenates into the alcohol phase, and decanting the d-limonene. Use of 20 parts alcohol/1 part oil is common. Use of this process still requires distillation to remove the alcohol for reuse and recover the folded oil, if the alcohol wash, itself, is not used as a flavor.

Distillation to manufacture 5-fold, 10-fold, and 20-fold oils does not result in a linear concentration of the flavor containing aldehydes of the oil, as part may be lost during the folding. Particularly, some octanal will be lost in the terpene fraction. A 10-fold oil may be in the range of 12–15% aldehydes, having lost 20–40% of the aldehydes of the original single-strength oil. Oils concentrated more than 20-fold are usually referred to as terpeneless oils because of the low concentrations of limonene, myrcene, and pinene. Storage, handling, and use of folded oils is similar to other oils, with the exception that folded oils have higher stability used in products such as candy or baked items, which may be heated.

11.3.2.2 *Adsorption*

Terpeneless and folded oils have been manufactured by adsorption with activated silica. Use of silica involves mixing the oil with the silica until the components are absorbed, washing the terpenes out of the silica with a nonpolar solvent, such as hexane, followed by elution of the oxygenates, pigments, and so forth with a polar solvent such as ethyl acetate (Kirchner and Miller, 1952). The solvent must be removed by distillation to recover the terpeneless oil, and the silica must either be regenerated or discarded, adding to the process cost. Flavor compounds may also be adsorbed from aqueous citrus essence by use of nonpolar polymeric materials such as styrene divinylbenzene (Bryan et al., 1977; Persson et al., 1990; Tseng et al., 1993).

11.3.2.3 *Poroplast Extraction*

An extraction technique, which involves solute transfer between an aqueous phase passing over a nonpolar support containing the oil on a hydrophobic surface has been described (Fleisher, 1994). This process functions by washing the soluble oxygenates from the absorbed oil with aqueous alcohol in a column. When the oil is exhausted of oxygenated compounds by the alcohol, fresh oil is introduced. No difficult emulsions are formed by this method; however, distillation of the aqueous alcohol phase is still necessary to recover the terpeneless oil.

11.3.2.4 Supercritical CO_2 Extraction Carbon dioxide, under supercritical conditions, can act as a solvent with properties dependent on the pressure and temperature, which can provide certain advantages over other separation processes. Thermally sensitive compounds can be separated at low temperatures, CO_2 is nontoxic and is easily removed from the product by releasing the pressure, and changes in pressure or temperature can change the solvent selectivity. CO_2 is also inert, nonflammable, available in high purity, and inexpensive. There have been some applications of CO_2 as an extraction solvent in the manner of alcohol washing for folding citrus oils (Japikse et al., 1987). For successful extractions, vapor pressures of oil components determine their solubility in supercritical solvents. The vapor pressure of limonene is approximately four times greater than that of linalool, an oxygenated terpene (Fig. 11.4). Since CO_2 has no dipole moment and a greater affinity for nonpolar solvents, terpenes with lower molecular weight and higher vapor pressures are more soluble in supercritical CO_2 than oxygenated flavor components (Temelli et al., 1988a,b).

Although the process is now commercial for citrus oils, extraction of the terpene fraction with CO_2 to fold the oil has its problems. The process is

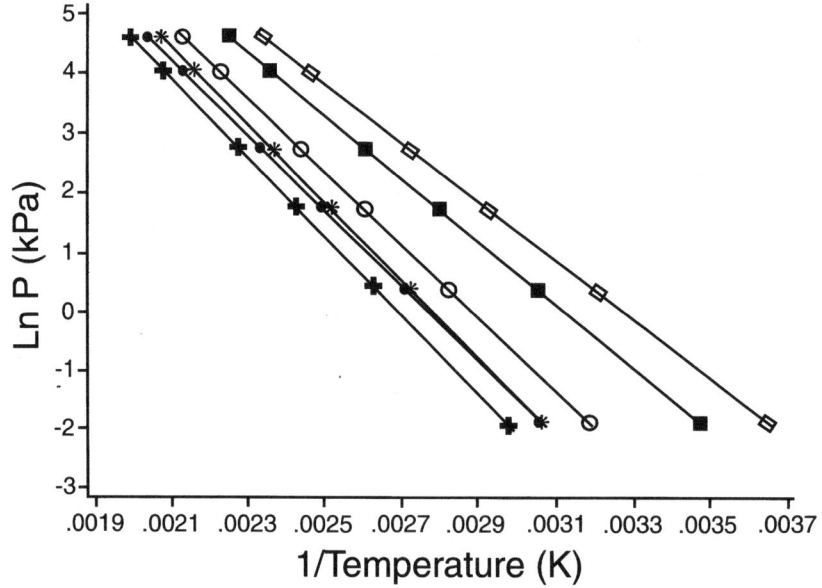

Figure 11.4 Vapor pressure vs. temperature graphs of some important components of cold-pressed orange oil: ($\square$) d-limonene, ($\blacklozenge$) α-pinene, myrcene, (○) linalool, (*) decanal, (●) α-terpineol, and (×) citral (plotted from data in CRC, 1986).

expensive, since vessels, columns, and equipment must withstand pressure up to 20 MPa. However, a major technical limitation is that above the critical point (31°C, 7.3 MPa), the conditions for highest selectivity of the solvent for the terpenes is near the critical temperature and not the conditions for the highest yields (Temelli et al., 1990). Because the folded oil is the product, not the extract, yield refers to terpenes extracted. There are some options, such as alcohol entrainment, which attempt to deal with this problem. One process involves use of silica to absorb oxygenates, CO_2 extraction of the terpenes, then changing pressure/temperature to extract oxygenates from the silica (Dugo et al., 1995). Folded oils produced by this method have excellent flavor and aroma qualities.

11.3.3 Encapsulated Oils

Beverage powders and certain dried foods require dried flavorings. Drying is not possible since citrus oil components are volatile liquids; thus the usual encapsulation procedure it to mix the oils with solvents, starch, gums, or corn syrup solids. This mixture may be extruded into a solvent (ethanol), then dried by evaporation (Swisher, 1962; Miller and Mutka, 1986). There are many processes and encapsulating agent combinations, with the objective of preparing good-flavored products stable to oxidative and storage changes (Anandaraman and Reineccius, 1986). Products extruded through dies are broken into small cylindrical pieces, which contain the flavor inside the matrix, protected from air and the environment of the product. The matrix itself is soluble in the product, releasing the flavor. Because the high surface area of encapsulated flavor particles may expose the oil to oxidation, an antioxidant such as BHA may be added during manufacture.

11.4 USE AND EVALUATION

The soft drink and beverage industries are the largest users of citrus oils for flavoring, followed by ice cream, cookies and desserts, confections, and chewing gum. The use level ranges from about 200 ppm in beverages to over 2000 ppm in chewing gum. World production of citrus essential oils has been estimated at greater than 50,000 mt, with 8000 mt used in soft drinks (Buchel, 1989). Production of the oil requires that the fruit be processed; thus schemes for estimating the amount of oil produced may use the values of total oil (Table 11.1) and fruit production/utilization statistics from the various growing areas (Chapter 1). It is also necessary to estimate the percent recovery from the fruit.

Evaluation of the product use level and potential flavor impact of the oils on consumers is ultimately performed by sensory evaluation at the research stage. Besides sensory techniques, analytical methods using gas chromatography/mass spectrometry (GC/MS) are useful objective techniques, which provide data related to process control as well as product stability. Books have been written related to analytical methodology of volatile flavors in foods and beverages. The present discussion will call attention to some recent important applications of chemistry of essential oils use in foods.

For application of citrus oils to drinks and beverages, preparation of emulsions of the various oils are necessary because of the low specific gravity and insolubility of the oils in the beverage matrix. Since this matrix is mostly water and sugars and the oils are terpenes, a weighting agent (ester gum) is mixed with the oil to increase the specific gravity closer to that of the beverage. The weighting agent is usually added to the oil (1:1) to increase the specific gravity to near that of water. Gum arabic is mixed with water separately for the aqueous phase of emulsions. Since the finished oil emulsion specific gravity is still much less than that of the beverage, which contains sugar, for the emulsion to be stable, the particle size must be reduced. A recent report has given details of beverage emulsions and shown the relationship between emulsion droplet size, the weighting agent, and films of gum or modified starch, conferring long-term stability of the flavor emulsion in the product (Tan, 1998).

Computers have allowed the application of complex statistics to evaluate the volatile composition of food package headspaces. Use of multivariate statistical packages has been demonstrated to be capable of making comparisons between drinks and pure juices, shelf-life changes, and the like based on analysis of the GC chromatogram of the headspace volatiles (Shaw and Moshonas, 1997). Over 40 volatile compounds in the headspaces of fresh and processed commercial orange juices were determined and principal component analysis used to determine juice characteristics (Moshonas and Shaw, 1997). It is very difficult to determine actual volatile concentrations important to a food's flavor, and to relate this information to the human sensation of taste. Partition of volatile compounds between vapor/liquid and solid phases, followed by release during eating, has been described (Taylor, 1998).

A simple technique has been developed to concentrate the volatiles in a package headspace for GC/MS analysis. This technique, solid-phase microextraction (SPME) involves placing a fiber coated with an absorbent specific for the compounds of interest in the headspace. These compounds absorb to the fiber kinetically, depending on the vapor pressure, temperature, and time. Once absorbed, the fiber is injected into the GC/MS, where the heat desorbs the compounds onto the column for analysis (Harmon, 1997). The convenience of this procedure for preparative and analytical analysis of volatile flavors has made this the method of choice in many instances.

The GC/MS column technology has developed to the high-tech state of being capable of separation and identification of the natural enantiomers of the essential oils. The usefulness of this technology cannot be underestimated, as it has long been known that chiral discrimination of some aroma molecules depends on the enantiomeric distribution of the compound. A number of odor perceptions of enantiomers of selected flavor compounds has been published, for example, R-(−)-carvone is herbaceous and dill-like and S-(+)-carvone is characterized as spearmint (Koppenhoefer et al., 1994). The use of GC/MS allows selecting the enantiomeric distribution of these molecules, many of which are specific in the natural product but may be altered in an adulterated flavor. Some characteristic enantiomeric ratio distributions of terpenes from citrus varieties have been published, which are useful for identifying specific oils (Mosandl et al., 1990). The reader can pursue the considerable scientific literature on this subject.

11.5 STORAGE AND HANDLING

Citrus oils should be shipped in full glass, tin-dipped, stainless steel, aluminum, or polymer-coated steel containers. Container size varies from small bottles for sample submission to 5-gal cans, 55-gal drums, or tank lots. Air should be excluded from containers by nitrogen flushing to minimize oxidation. The oil should be dry, as the presence of water will allow slow hydrolysis and formation of products such as α-terpineol. Storage at air-conditioning temperatures is adequate for most oils as wax will not precipitate under this condition. However, cold-room storage may be used for certain essences and essence oils, which do not contain the natural antioxidants of the cold-pressed oils.

Uniform color and flavor quality of oils is maintained by blending and large quantity bulking in tanks. These tanks vary in size depending on the specific oil and its demand volume but may vary from a few hundred to as much as 5000–10,000 gal. Blending of raw product at the citrus plant may help processors meet certain uniformity requirements at the point of sale. Also, bulk samples require fewer quality tests at the laboratory level. Sampling should be from well-mixed tanks or drums to represent the actual quality of the lot, ensuring the quality to the buyer.

Certain oils may be stored in specialized containers. For example, lime and lemon oil may be stored and shipped in tin-dipped drums, which adds expense to the value of these oils. Storage of some oils in polymer-coated cans and drums may be satisfactory for a short time, but many polymers (particularly polyethylene) are incompatible with the terpenes of the oil. Thus, for long-term storage, suitable containers must be used.

REFERENCES

Anandaraman, S. and Reineccius, G. A. (1986). Stability of encapsulated orange peel oil, *Food Technol.* **40**(1):88–93.

Bartholomew, E. T. and Sinclair, W. B. (1946). Factors influencing the volatile oil content of the peel of immature and mature oranges, *Plant Physiol.* **21**:319–331.

Bennett, A. H. (1934). Concentrated citrus oils, *Perf. Ess. Oil Record* **April**:111–112.

Braddock, R. J. and Kesterson, J. W. (1976). Quantitative analysis of aldehydes, esters, alcohols and acids from citrus oils, *J. Food Sci.* **41**(5):1007–1010.

Braddock, R. J. and Miller, W. M. (1982). Mass flow and energy use during orange peel oil recovery, *J. Food Sci.* **47**(6):2008–2010.

Braddock, R. J., Sadler, G. D., and Chen, C. S. (1991). Reverse osmosis concentration of aqueous-phase citrus juice essence, *J. Food Sci.* **56**(4):1027–1029.

Braddock, R. J., Temelli, F., and Cadwallader, K. R. (1986). Citrus essential oils—a dossier for material safety data sheets, *Food Technol.* **40**(11):114–116.

Braverman, J. B. S. (1949). *Citrus Products; Chemical Composition and Chemical Technology*, Interscience, New York.

Brent, J. A., DuBois, C. W., and Huffman, C. F. (1966). Essence recovery, U.S. Pat. 3,248,233.

Bryan, W. L., Lund, E. D., and Wagner, C. J., Jr. (1977). Adsorption of flavor components from aqueous orange peel aroma solutions, *Ind. Eng. Chem., Prod. Res. Dev.* **16**(3):257–261.

Buchel, J. A. (1989). Flavoring with citrus oils, *Perf. Flavorist* **14**(1):22–26.

Bushman, R. C. (1972). Apparatus for extracting citrus peel oil from whole fruit, U.S. Pat. 3,707,176.

Cadwallader, K. R. and Xu, Y. (1994). Analysis of volatile components in fresh grapefruit juice by purge and trap gas chromatography, *J. Agric. Food Chem.* **42**(3):782–784.

Chemical Market Reporter (1998). Chemical prices, *Chem. Market Report.* **254**(7):26–35.

Clark, B. C., Jr. and Chamblee, T. S. (1992). "Acid-catalyzed reactions of citrus oils and other terpene-containing flavors." In *Off-Flavors in Foods and Beverages, Developments in Food Science*, Vol. 28, G. Charalambous, Ed., Elsevier Science, Amsterdam, pp. 229–285.

CRC (1986). "Vapor pressures. Organic compounds less than 1 atmosphere." In *CRC Handbook of Chemistry and Physics*, 67th ed., CRC Press, Boca Raton, FL.

Demole, E. and Enggist, P. (1983). Further investigation of grapefruit juice flavor components (*Citrus paradisi* Macfayden). Valencane- and eudesmane-type sesquiterpene ketones, *Helv. Chim. Acta* **66**(5)Nr.131:1381–1391.

Demole, E., Enggist, P., and Ohloff, G. (1982). 1-*p*-Menthene-8-thiol: A powerful flavor impact constituent of grapefruit juice (*Citrus paradisi* Macfayden), *Helv. Chim Acta* **65**(6)Nr.176:1785–1794.

Distelkamp, A. (1962). Centrifuges in the citrus industry, *Trans. Citrus Eng. Conf.* **8**: 41–50.

Dugo, P., Mondello, L., Bartle, K. D., Clifford, A. A., and Breen, D. G. (1995). Deterpenation of sweet orange and lemon essential oils with supercritical carbon dioxide using silica gel as an adsorbent, *Flavor Fragrance J.* **10**:51–58.

Ferguson, R. R. (1980). Procedures for determining balances in oil mill, *Citrus Ind. Mag.* **61**(12):27–36.

Fleisher, A. (1994). Citrus hydrocarbon-free essential oils, *Perf. Flavorist* **19**(1):11–12, 14–15.

Food Chemicals Codex (1996). *Monographs*, 4th edition, National Academy Press, Washington, D.C.

FMC (1998). By-product capabilities, http://www.fmcfoodtech.com.

Goodrich, R. M., Braddock, R. J., Parish, M. E., and Sims, C. A. (1998). Bioconversion of citrus aroma compounds by *Pichia pastoris*, *J. Food Sci.* **63**(3):445–449.

Guenther, E. (1949). *Individual Essential Oils of the Plant Families Rutaceae and Labiatae*, Vols. 1 and 3, *The Essential Oils*, Van Nostrand, New York.

Harmon, A. D. (1997). Solid-phase microextraction for the analysis of flavors. In *Techniques for Analyzing Food Aroma*, R. Marsili, Ed., Marcel Dekker, New York, Chap. 4, pp. 81–112.

Hendrickson, R., Kesterson, J. W., and Cohen, M. (1970). Effect of budwood selection and rootstock on the peel oil content of Valencia oranges, *Proc. FL State Hort. Soc.* **83**:259–262.

Hendrix, D. L., Nagy, S., Balaban, M. O., and Crandall, P. G. (1992). Effects of temperature on wax yields and composition during winterization of commercial cold-pressed oil, *Proc. FL State Hort. Soc.* **105**:146–148.

Hood, S. C. and Russell, G. A. (1916). The production of sweet orange oil and a new machine for peeling citrus fruits, USDA Bull. No. 399, Washington, D.C.

Huet, R. (1991). Les huiles essentielles d'agrumes, *Fruits* **46**(5):551–576.

Janes, J. F., Marr, L. M., Unwin, N., Banthorpe, D. V., and Yusuf, A. (1993). Reaction of monoterpenoids with hydrogen sulfide to form thiols and epi-sulfides of potential organoleptic significance, *Flavor Fragrance J.* **8**:289–294.

Japikse, C. H., Van Brocklin, L. P., Hembree, J. A., Kitts, R. R., and Meece, D. R. (1987). Process for the production of citrus flavor and aroma compositions, U.S. Pat. 4,647,466.

Johnson, J. D. and Vora, J. D. (1983). Natural citrus essences, *Food Technol.* **37**(12): 92–93, 97.

Kesterson, J. W. and Braddock, R. J. (1975). Total peel oil content of the major Florida citrus cultivars, *J. Food Sci.* **40**(5):931–933.

Kesterson, J. W. and Braddock, R. J. (1976). By-products and specialty products of Florida citrus, FL Agr. Exp. Sta. Tech. Bull. No. 784, University of Florida, Gainesville, FL.

Kesterson, J. W. and Braddock, R. J. (1977). Influence of cultural practices, scion and rootstock selection on citrus peel oil production, *Proc. Int. Soc. Citriculture* **3**:734–736.

Kesterson, J. W. and Hendrickson, R. (1962). The composition of Valencia orange oil as related to fruit maturity, *Am. Perf. Cosmetics* **77**(12):21–24.

Kesterson, J. W. and Hendrickson, R. (1967). Curing Florida grapefruit oils, *Am. Perf. Cosmetics* **82**(1):37–40.

Kesterson, J. W. and McDuff, O. R. (1948). Florida citrus oils. Commercial production methods and properties, FL Agr. Exp. Sta. Tech. Bull. No. 452, University of Florida, Gainesvill, FL.

Kesterson, J. W., Hendrickson, R., Seiler, R. R., Huffman, C. E., Brent, J. A., and Griffiths, J. T. (1965). Nootkatone content of expressed Duncan grapefruit oil as related to fruit maturity, *Am. Perf. Cosmetics* **80**(12):29–31.

Kesterson, J. W., Hendrickson, R., and Braddock, R. J. (1971). Florida citrus oils, FL Agr. Exp. Sta. Tech. Bull. No. 749, University of Florida, Gainesville, FL.

Kesterson, J. W., Braddock, R. J., Koo, R. C. J., and Reese, R. L. (1977). Nitrogen and potassium fertilization as related to the yield of peel oil from pineapple oranges, *J. Am. Soc. Hort Soc.* **102**(1):3–4.

Kirchner, J. G. and Miller, J. M. (1952). Preparation of terpeneless essential oils, a chromatographic process, *Ind. Eng. Chem.* **44**(2):318–321.

Koppenhoefer, B., Behnisch, R., Epperlein, U., Holzschuh, H., Bernreuther, A., Piras, P., and Roussel, C. (1994). Enantiomeric odor differences and gas chromatographic properties of flavors and fragrances. *Perf. Flavorist* **19**(Sept./Oct.):1–14.

Lawrence, B. M. (1992). Progress in essential oils. Tangerine oil. Mandarin oil. *Perf. Flavorist* **17**(July/Aug.):50–54.

Maarse, H. and Visscher, C. A. (1989). "Citrus fruits. Product 5." In *Volatile Compounds in Food. Quantitative and Qualitative Data*, TNO-CIVO Food Analysis Institute, Zeist, The Netherlands.

Markley, K. S., Nelson, E. K., and Sherman, M. S. (1937). Some waxlike constituents from expressed oil from the peel of Florida grapefruit, *J. Biol. Chem.* **118**:433–441.

Matthews, R. F. and Braddock, R. J. (1987). Recovery and applications of essential oils from oranges. *Food Technol.* **41**(1):57–61.

McDonald, R. E. and Hillebrand, B. M. (1980). Physical and chemical characteristics of lemons from several countries, *J. Am. Soc. Hort. Soc.* **105**(1):135–141.

McKinney, Jr., J. E. (1981). Apparatus for extracting citrus peel oil from whole fruit, U.S. Pat. 4,248,142.

Miller, D. H. and Mutka, J. R. (1986). Preparation of solid essential oil flavor composition, U.S. Pat. 4,610,890.

Mookherjee, B. D., Chant, B. J., Evers, W. J., Wilson, R. A., Zampino, M. J., and Vock, M. H. (1985). Process for preparing mixtures containing 8,9-epithio-1-*p*-menthene, U.S. Pat. 4,536,583.

Mosandl, A., Hener, U., Kreis, P., and Schmarr, H. G. (1990). Enantiomeric distribution of α-pinene, β-pinene and limonene in essential oils and extracts. Part 1. *Rutaceae* and *Gramineae, Flav. Fragrance J.* **5**:193–199.

Moshonas, M. G. and Shaw, P. E. (1974). Quantitative and qualitative analysis of tangerine peel oil, *J. Agric. Food Chem.* **22**(2):282–284.

Moshonas, M. G. and Shaw, P. E. (1997). Dynamic headspace gas chromatography combined with multivariate analysis to classify fresh and processed orange juices, *J. Essent. Oil Res.* **9**:133–139.

Moshonas, M. G., Shaw, P. E., and Veldhuis, M. K. (1972). Analysis of volatile constituents from Meyer lemon oil, *J. Agric. Food Chem.* **20**(4):751–752.

Naves, Y. R. (1947). Contribution to our knowledge of the aldehyde fraction of essential oil of sweet orange, *Perf. Ess. Oil Rec.* **83**(8):295–298, 320.

Nigg, H. N., Nordby, H. E., Beier, R. C., Dillman, A, Macias, C., and Hansen, R. C. (1993). Phototoxic coumarins in limes, *Food Chem. Toxicol.* **31**(5):331–335.

Nordby, H. E. and McDonald, R. E. (1990). Squalene in grapefruit wax as a possible natural protectant against chilling injury, *Lipids* **25**(12):807–810.

Nordby, H. E. and McDonald, R. E. (1994). Friedelin, the major component of grapefruit epicuticular wax, *J. Agric. Food Chem.* **42**:708–713.

Persson, L. T., Van Brocklin, L. P., Morrison, L. R., Jr., Smith, C. A., and Meece, D. R. (1990). Process for making improved citrus aqueous essence and product produced therefrom, U.S. Pat. 4,970,085.

Poore, H. D. (1932). Analyses and composition of California lemon and orange oils, Tech. Bull. 241, USDA, Washington, D.C.

Raymond, W. R. (1984). Process for the generation of acetaldehyde from ethanol, U.S. Pat. 4,481,292.

Redd, J. B., Shaw, P. E., Hendrix, C. M., Jr., and Hendrix, D. L. (1996). "Flavor concentration systems." In *Quality Control Manual for Citrus Processing Plants*, Vol. III, AgScience, Auburndale, FL, Chap. 2.

Schieberle, P. and Grosch, W. (1988). Identification of potent flavor compounds in an aqueous lemon oil/citric acid emulsion, *J. Agric. Food Chem.* **36**:797–800.

Shaw, P. E. (1991). "Fruits II." In *Volatile Compounds in Foods and Beverages*. H. Maarse, Ed., Marcel Dekker, New York, Chap. 9.

Shaw, P. E. (1996). "Volatile components important to citrus flavors." In *Quality Control Manual for Citrus Processing Plants*, Vol. 3, J. B. Redd, P. E. Shaw, C. M. Hendrix, Jr., and D. L. Hendrix, Eds., AgScience, Auburndale, FL, Chap. 4, pp. 134–172.

Shaw, P. E. and Moshonas, M. G. (1997). Quantification of volatile constituents in orange juice drinks and its use for comparison with pure juices by multivariate analysis, *Lebensm.-Wiss. u.-Technol.* **30**:497–501.

Sinclair, W. B. (1984). *The Biochemistry and Physiology of the Lemon and Other Citrus Fruits*, University of California Press, Oakland, CA.

Steger, E. S. (1979). Peel oil recovery by recycling centrifuge effluent, *Trans. Citrus Eng. Conf.* **25**:46–60.

Steger, E. (1981). Effect of processing on aldehyde content of Valencia oranges, *Citrus Ind. Mag.* **62**(8):5, 7–8, 12, 14.

Swisher, H. E. (1962). Solid essential oil flavoring compositions, U.S. Pat. 3,041,180.

Tan, C. T. (1998). "Beverage flavor emulsion—a form of emulsion liquid membrane microencapsulation." *Food Flavors: Formation, Analysis and Packaging Influences*, E. T. Contis, et al., Eds., Elsevier Science, Holland, pp. 29–42.

Taylor, A. J. (1998). Physical chemistry of flavor, *Int. J. Food Sci & Technol.* **33**:53–62.

Temelli, F., Braddock, R. J., Chen, C. S., and Nagy, S. (1988a). "Supercritical carbon dioxide extraction of terpenes from orange essential oil." In *Supercritical Fluid Extraction and Chromatography: Techniques and Applications*, ACS Symp. Ser. No. 366, B. A. Charpentier and M. R. Sevenants, Eds., American Chemical Society, Washington, D.C., Chap. 6, pp. 109–126.

Temelli, F., Chen, C. S., and Braddock, R. J. (1988b). Supercritical fluid extraction in citrus oil processing, *Food Technol.* **42**(6):145–150.

Temelli, F., O'Connell, J. P., Chen, C. S., and Braddock, R. J. (1990). Thermodynamic analysis of supercritical carbon dioxide extraction of terpenes from cold-pressed orange oil, *Ind. Eng. Chem. Res.* **29**(4):618–624.

Thrush, R. E. (1964). Application of centrifuges in the recovery of citrus flavor fractions, *Trans. Citrus Eng. Conf.* **10**:49–63.

Tseng, D. J., Matthews, R. F., Gregory, J. F., Wei, C. I., and Littell, R. C. (1993). Sorption of ethyl butyrate and octanal constituents of orange essence by polymeric adsorbents, *J. Food Sci.* **58**(4):801–804.

Vora, J. D., Matthews, R. F., Crandall, P. G., and Cook, R. (1983). Preparation and chemical composition of orange oil concentrates, *J. Food Sci.* **48**(4):1197–1199.

Waters, R. (1993). Coldpressed citrus oil recovery, *Trans. Citrus Eng. Conf.* **39**:28–48.

Waters, R. D., Kesterson, J. W., and Braddock, R. J. (1976). Method for determining the α-tocopherol content of citrus essential oils, *J. Food Sci.* **41**:370–371.

CHAPTER 12

d-LIMONENE

The chemical d-limonene has been a stable citrus by-product for over 50 years. Originally called citrus stripper oil, d-limonene is recovered by distillation from press liquor, waste streams, peel oil emulsions, and from folding citrus oils. It is the major constituent of the oil in the peel of all edible citrus fruits, a member of the chemical family of cyclic monoterpenes, and has the formula $C_{10}H_{16}$. World production of d-limonene is estimated to be between 50,000 and 75,000 mt (110 and 165 million lb), most from Brazil and Florida. Its uses include production of adhesive resins, flavors, solvents, and degreasing agents.

12.1 RECOVERY

The recovery of d-limonene as a product is intimately tied to cold-pressed oil recovery from the peel. Some processors do not recover cold-pressed oil from early and midseason fruit and some recover none at all, greatly affecting the amount in the peel at the feed mill. In any event, the amount recovered may be no more than the total for each variety of fruit, discussed in Chapter 11 (Table 11.1). In other instances, because the world supply of cold-pressed orange oil occasionally exceeds the demand, the price of the lower aldehyde products falls to near the value of d-limonene. Then, although cold-pressed oil is more expensive to recover, it may be sold as d-limonene.

12.1.1 Press Liquor

The peel sent to the feed mill contains the remainder of the oils, which were not extracted in the cold-pressed oil recovery process. After liming and pressing, the press liquor is the source for d-limonene recovered during molasses concentration in the waste heat evaporator (Chapter 10). The concentration of d-limonene in the press liquor is influenced by many process variables but is usually in the range of 0.1–0.5% (v/v). Press liquor water vapor from the first effect tubes is enriched in d-limonene, condensed and passes to a tank at the bottom of the evaporator (Odio, 1993). This tank contains baffles and material for increasing surface area to break the emulsion of d-limonene and condensate. This is a continuous process allowing the d-limonene to float and be recovered by decanting, where it is sent to a storage tank. The product oil is primarily composed of the chemical d-limonene, which is the name by which it is traded. An older synonym, stripper oil, derives from the steam stripping process used in conjunction with the traditional 72 °Brix molasses process (Schulz, 1972).

The decant tank underflow is usually sent to waste treatment or to a spray field, which may result in other problems. First, because of the way the press liquor is handled, allowing partial fermentation of some sugars to form alcohol, the aqueous decant discharge may be enriched in ethanol. Second, the ethanol concentration of this condensate may be as high as 2 or 3%, increasing its solubility to d-limonene and resulting in lost yield. When this stream is sent to waste treatment, the additional d-limonene may cause problems, as it is very toxic to microbes used in such processes (Graumlich, 1983). Recovery of d-limonene from decant tank and evaporator condensate water by reverse osmosis has been studied (Braddock and Adams, 1984). Reaction of d-limonene with certain membrane polymers indicated that polysulfone hollow fibers were unsuitable. However, Teflon membranes were capable of rejecting d-limonene down to 0.003% concentration (Braddock, 1982).

12.1.2 Oil Emulsion

When cold-pressed oil is not recovered by centrifugation, the d-limonene may be recovered from the oil–water emulsion by steam distillation. In this process, the d-limonene is insoluble in the water and is distilled over by the steam at reduced temperature. This is not a typical fractional distillation, as practiced when distilling cold-pressed oil, but rather a bulk separation of the volatile compounds in the condensed vapor. Applied to d-limonene, the ratio of the partial pressures of water (P_w) and d-limonene (P_l) is the ratio of the mole fraction of water (M_w) and limonene (M_l), that is, $P_l/P_w = M_l/M_w$. This ratio is an application of Raoult's law for the minimum d-limonene to water mass (or volume) ratio achievable at a given system pressure. Because water/d-limonene systems are saturated (immiscible), due to the poor solubility of d-limonene, the vapor concentration of d-limonene is constant at equilibrium. This should allow complete recovery of the d-limonene by steam distillation at the ratio of water to d-limonene vapor pressures.

The mass ratio of water/d-limonene to be distilled for recovering d-limonene has been studied and at atmospheric pressure is in the range of 8–10/1. Figure 12.1 indicates the type of curves obtained at atmospheric pressure distillation

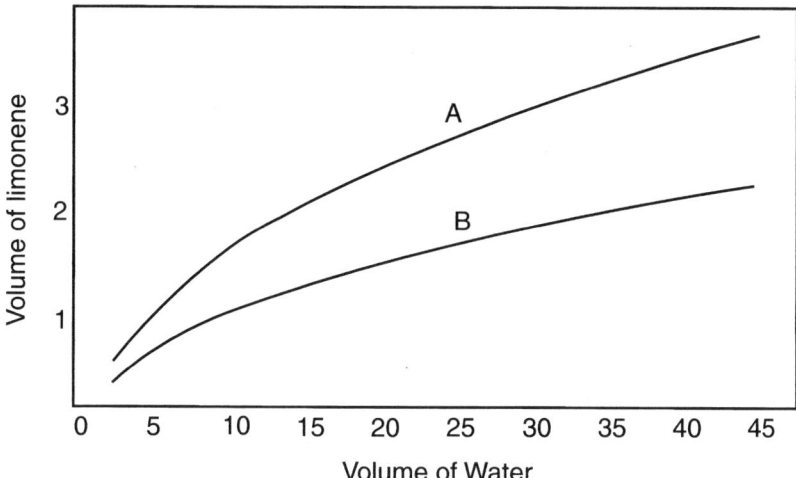

Figure 12.1   Relative volume recovery of d-limonene per volume of water for atmospheric pressure distillation of (A) 3% (v/v) d-limonene in water and (B) 3% peel oil emulsion.

of mixtures of d-limonene and water (Fig. 12.1*a*) and oil emulsion (Fig. 12.1*b*). Unfortunately, in the situation with actual oil emulsions (12.1*b*), considerably more water must be distilled to recover a given quantity of d-limonene because the d-limonene is adsorbed by the pulp particles in the emulsion. If the distillation is performed in a continuous manner under pressure, the ratio of water removed per amount of d-limonene recovered can be reduced to 3–4/1 (Gerow, 1974). One problem occurs during distillation of d-limonene above atmospheric pressure. Once the steam temperature passes above 120°C (250°F), recovery of d-limonene decreases, due to formation of some water-soluble alcohols and epoxides, which are not decanted with the d-limonene phase above the condensate.

12.1.3   Folded Oil

d-Limonene is also a by-product of the folding of the various cold-pressed and essence oils. Because of the effect of temperature degrading the oil components, these distillations are performed under reduced pressure. Very efficient flavor house stills are capable of fractionating cuts of very pure d-limonene to over 99% purity. These distillations generally begin at lower vacuum to recover the low boiling components, increasing the vacuum to recover the d-limonene. Careful attention to the still is required to separate the d-limonene from close boiling fractional impurities such as myrcene and octanal. If additional purity is required, distillation over dilute NaOH or a carbonyl-adduct agent (e.g., hydroxylamine hydrochloride) can remove aldehydes producing commercial d-limonene of 99.5% purity. Activated silica may also be used as an adsorbent to remove carbonyls and oxygenated impurities. Odiferous substances may also be removed by washing the d-limonene with an aqueous oxidizing agent, such as hypochlorite (Kruger et al., 1993). After purification, d-limonene will rapidly react with oxygen in the air and, if not protected, will revert to the crude d-limonene of 95–96% purity.

12.2 STORAGE AND TRANSPORTATION

12.2.1 Storage

All citrus oils used for flavors should be stored in the absence of air at cool temperatures. Since d-limonene may have industrial uses unrelated to flavor, bulk storage in inert atmosphere may be in iron tanks at ambient conditions, although off-flavors and some yellowish color may develop. Storage of d-limonene in iron tanks helps prevent peroxide buildup; however, stainless steel tanks are now in use, with provisions for nitrogen blanketing. Smaller containers are used for shipping, which may range from 100-mL aluminum or glass sample bottles to 18-L (5-gal) or 208-L (55-gal) drums with inert epoxy-type liners. Drums not blanketed with nitrogen should be filled completely through the bung to minimize the headspace air, generally requiring 182 kg (400 lb) of oil. Elimination of air reduces oxidation to carvone and carveol and other off-odors. The d-limonene should also be dry, as water may react with exposed iron in lined drums, which causes discoloration. Other contaminants in the container may also deteriorate the d-limonene, if the container has been used and not properly cleaned. Cold-pressed oils in drums have been known to develop a skunky odor and blacken due to reaction of flavonoids in the oil with iron in the presence of small amounts of water.

At elevated storage temperatures, d-limonene in the presence of air can polymerize, decreasing the optical rotation and thus the purity. This reaction may also be free-radical catalyzed, aided by sunlight exposure. Commercial storage may warrant addition of a suitable antioxidant, such as 50–200 ppm BHT. This antioxidant is volatile and may be removed for many industrial limonene applications. Nitrogen blanketing in large tanks performed to minimize oxidation may be done by maintaining a small positive pressure, allowing filling and removal of d-limonene.

12.2.2 Transportation and Safety

In the United States, transportation rules and regulations for d-limonene and other citrus oils are complex and involved. These oils have hazardous properties, which meet Occupational Safety and Health Administration (OSHA) regulations requiring Material Safety Data Sheets (MSDS). In-depth MSDS property data for d-limonene has been compiled and published to aid those seeking to comply with the regulations (Braddock et al., 1986). One of the MSDS hazard data properties, flashpoint (d-limonene = 115°F, 46°C), triggers the Department of Transportation (DOT) Hazardous Materials Regulations (HMR), HM-181. Also, if d-limonene (or oil) is spilled or discharged as a waste, the Comprehensive Environmental Response, Compensation and Liability Act (CERCLA) is applied according to a chemical's flashpoint. The U.S. Code of Federal Regulations (CFR) contains the information necessary for reference matters on these issues as follows: 49 CFR § 172.101, 49 CFR § 173.50 (f) (1), 49 CFR § 173.120, 49 CFR § 173.300, 49 CFR § 173.115, and 40 CFR § 261.210. Flashpoints of the various citrus oils of commerce are published by the Essential Oil Association (FMA, 1985).

The Hazardous Materials Table (HMT) found in 49 CFR § 172.101 is the handbook for applying the HMR to DOT shipments of d-limonene and citrus oils. There is no class name listed in the HMT into which citrus oils are specifically classified (e.g., C.P. Orange Oil). However, there are several generic class names that apply to d-limonene and various citrus oils. The following classes in the HMR are listed: UN 1993 Combustible Liquid, n.o.s. (USA only), UN 1993 Flammable Liquid, n.o.s., UN 1197 Extracts, Flavoring, Liquid (International Class 3 Flammable Liquid), UN 2319 Terpene Hydrocarbons, n.o.s. All citrus oils are considered either flammable or combustible liquids with a flammability rating = 2 (NFPA, 1975).

12.3 PROPERTIES

12.3.1 Chemical

12.3.1.1 Nomenclature The compound d-limonene has the chemical name (4R)-(+)-4-isopropenyl-1-methylcyclohexene, Chemical Abstracts Registry Number 5989-27-5. *Stereoisomers* are compounds of identical structure but differing in the arrangement of their atoms in three-dimensional space. On paper, a method describing atom location for molecules, R-CHX-R′, is to draw from top-to-bottom the main chain, R-C-R′. If X is on the right side, the molecule is called D; if X is on the left it is L. Modern chemists have replaced the D, L system with a system where atoms are arranged *a*, *b*, *c*, *d* around a central carbon atom. The molecule is then viewed remote from *d*. If atomic number of the order *a*, *b*, *c* (according to decreasing atomic number) traces clockwise (right) around the carbon atom, the configuration is R, counterclockwise (left, or sinister), configuration is S. These spatial relationships affect physical properties related to geometry of the molecules.

The most important measurable physical property of each isomer of stereoisomers, *enantiomers*, is the ability to rotate a plane of polarized light to the right (+) or left (−) by a certain number of degrees (α). To the right, *dextrorotatory*, was historically referred to as d, and left as *levorotatory*, l. Hence, this describes the common use of the older designation of d-limonene for (+)-limonene. Equal mixtures of (+) and (−) enantiomers are called *racemates*. As an example, at 25°C, crude citrus (+)-limonene, (4R)-(+)-4-isopropenyl-1-methylcyclohexene, can rotate light at least 96° to the right ($\alpha = +96°$). This number increases with increasing purity ($\alpha = +126°$) for the purest d-limonene. Another example, dipentene (limonene from pine trees) is a *racemic* mixture of d- and l-limonene and does not rotate light ($\alpha = 0$). Sometimes in the trade, one sees written the upper case D (i.e., D-limonene). Since this refers to the spatial configuration, and not the optical rotation of light, it is incorrect.

12.3.1.2 Analysis For most flavor and purity applications, d-limonene is commonly analyzed by GC/MS. However, wet chemical analysis still is used for certain applications, particularly by the industry using citrus d-limonene as a chemical starting material for synthesis of adhesive resins. Some of the useful properties of crude d-limonene are listed in Table 12.1. Analytical methods for properties such as optical rotation, refractive index, aldehydes, and specific gravity are the same as for cold-pressed orange oil discussed in Chapter 11.

The most common quantitative wet chemical analysis for d-limonene, referred to as the Scott test, is based on an edible oil bromination reaction to determine the number of fatty acid double bonds. The test is commonly used to measure the peel oil content of citrus juices by adding a small amount of isopropanol to the juice, co-distilling the alcohol and d-limonene from the juice, adding acid and indicator (methyl orange), followed by titration with the KBr-$KBrO_3$ reagent to the end point (color disappearance). The official procedure for using the Scott test, referred to as Recoverable Oil Test, for various citrus juices and concentrate has been published (AOAC, 1995). The chemistry of this important reaction is illustrated in Figure 12.2, with the release of Br_2 and formation of limonene tetrabromide. This estimation of the oil in orange, grapefruit, and tangerine juice based on d-limonene is accurate since the oil from these fruit is >95% d-limonene. A minor adjustment may be made for lemon and lime juices, whose oils have 5% α-pinene and 4% citral. Because α-pinene has only one double bond, under conditions for the reaction, it undergoes rearrangement to consume bromine at the same rate as d-limonene; however, citral consumes bromine at half the d-limonene rate (Scott and Veldhuis, 1966). Precision of this method for juices has been determined to be near 10 ppm d-limonene.

The Scott test also may be applied to analysis of the d-limonene content of feed mill peel residue, press cake, press liquor, and whole fruit. For this application, a suitable sample size must be adequately disintegrated in enough liquid in a blender to take a uniform sample for the distillation and subsequent titration. It is suggested that samples of wet peel residue, press cake, and pellets be taken based on 100 g of dry matter (0% moisture). The following procedure has been used in the author's laboratory with consistent results.

Sample size:

$$\text{Wet peel residue (20\% solids)} = 100 \text{ g} \div 0.2 = 500 \text{ g}$$

$$\text{Press cake (40\% solids)} = 100 \text{ g} \div 0.4 = 250 \text{ g}$$

$$\text{Pellets (90\% solids)} = 100 \text{ g} \div 0.9 = 111 \text{ g}$$

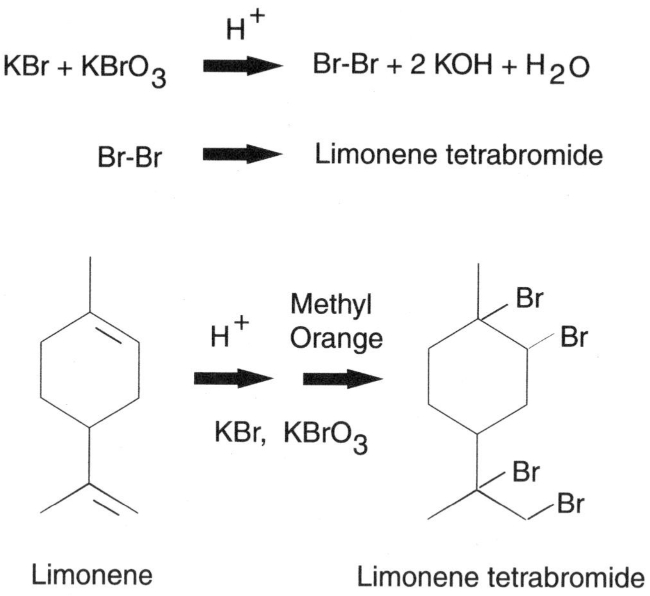
H+
KBr + KBrO3
Br-Br + 2 KOH + H2O
Br-Br
Limonene tetrabromide
Methyl
Orange
H+
KBr, KBrO3
Br
Br
Br
Br
Limonene
Limonene tetrabromide

Figure 12.2 Chemical reaction sequence of the reaction of d-limonene with bromine in the Scott test for recoverable oil.

All samples are brought to a mass of 3000 g with water for blending. If the blender will not hold 3000 g, smaller masses can be used (e.g., divide by 5), although the statistical uniformity of the sampling procedure favors use of larger samples. A critical issue is that enough water must be used to have a uniform liquid sample for the Scott oil procedure.

Procedure for wet residue:

Start with 500 g of wet peel residue.
Bring to 3000 g with water.
Blend in large blender 2 min at medium speed + 1 min at high speed.
Take duplicate 25-g samples for Scott oil analysis (25 g in 300 mL round distillation flask, 4 boiling chips + 25 mL 2-propanol).
Distill as per Scott oil analysis for juice.
Use 0.025 or 0.10 N bromide-bromate to titrate the distillate.

A sample calculation to express oil (lb oil/ton peel) follows:

$$11.3 \text{ mL titrant (0.025 N)/25 g sample} \times 3000 \text{ g sample/500 g peel}$$

$$\times\ 0.001 \text{ mL oil/mL titrant} \times 0.84 \text{ g oil/mL oil} \times 454 \text{ g peel/lb peel}$$

$$\times\ 1 \text{ lb oil/454 g oil} \times 2000 \text{ lb peel/ton peel} = 4.6 \text{ lb oil/ton peel}$$

12.3.1.3 Chemical Constants The numerable applications of d-limonene used in flavorings, industrial solvents, heat exchange fluids, and chemical syntheses require many physical and chemical constants, besides those listed in Table 12.1. The data compiled and listed in Table 12.2 from various literature sources are useful for the many applications of d-limonene in industry.

12.3.2 Environmental

It is probably true that all green plants in nature produce some limonene through their biochemical metabolism. The molecular structure and chemical reactivity of d-limonene implies that this chemical should degrade in the environment. Even chlorinated hydrocarbons can be degraded in certain microbial pathways. The specifics of such degradation in a natural environment is not well defined in situations of a large d-limonene spill, or disposal of excessive amounts to a sewage or waste treatment system. A label claim for biodegradability might not be so apparent, without definition of specific conditions.

Inherent biodegradation of organic substances can be tested (40 CFR § 796.334 Ch. 1, 7-1-92). Under this test, a substance with greater than 70% loss of dissolved carbon is evidence of ultimate biodegradability. The time period may be several months for some compounds. A number of limitations applying this test to d-limonene should be considered, for example, water solubility, vapor pressure, and inhibition of certain degradative microorganisms. As described, the test is applicable to water-soluble, nonvolatile organic chemicals that do not inhibit bacteria at test concentrations. d-Limonene has limited water solubility (13.8 mg/L at 25°C), is volatile (Table 12.2), and can inhibit certain

TABLE 12.1 Properties of Crude d-Limonene from Waste Heat Evaporator Recovery

Property	Value
Specific gravity (20°C)	0.84
Refractive index (20°C)	1.471
Optical rotation (20°C)	+96.7
Flash point (°C)(°F)	46 (115)
Boiling point (°C)(°F)	165 (330)
d-Limonene (%, GC)	94.6
High boilers (%)	2.3
Myrcene (%)	1.8
α-Pinene (%)	0.6
Sabinene (%)	0.4
Aldehydes (%)[a]	0.4
Low boilers (%)	0.1
Color	Colorless to pale yellow

[a]Expressed as octanal.

TABLE 12.2 Chemical and Physical Properties of d-Limonene

Property	Value	Property	Value
Commercial purity (%)	94 (crude)–99.6 (high purity)	Iodine No.	79
Color	Colorless	Acid value	0.03–0.4
Odor	Citrus, odorless at high purity	Kauri Butanol No.	59–63
Molecular wt.[a] (g/mol)	136.23	Peroxide No.	2–5
Boiling pt.[a] (°C, 763 mm Hg)	178 (high purity)	Optical rotation[a]	+96 (crude)
Freezing pt.[a] (°C)	−96.9	(α, 25°C)	+126 (pure)
Vapor pressure[b]		Solubility parameter[a] (δ)	8.2H
(mm Hg at 0, 25°C)	0.41, 2.1	Dipole moment (D)	0.61
Heat of vaporization[b,c]		Solubility[h] (mg/L at 25°C)	
(cal/g at 165, 25°C)	69.5, 84.4	In pure water	13.8
Specific heat (cal/g/°C)	0.438	In 20% sucrose	13.5
Heat of formation[c]		Viscosity[i] (cP)	
(kcal/mol at 25°C)	−13.0	at −50°C	3.5
Heat of combustion[d]		at 0°C	1.5
(kcal/mol)	1471	at 25°C	0.9
Flammable limits[e]		at 50°C	0.7
Autoignition (°C)	237	at 178°C	0.25
Lower explosive limit	0.7% vol. at 150°C	Surface tension[d] (dyn/cm)	
Upper explosive limit	6.1% vol. at 262°C	at 11°C	28.5
Flash point[f] (°C, open cup)	46 (crude)–49 (high purity)	at 25°C	27.0
Dielectric constant[g]		at 90°C	21.2
10 MHz at 20°C	2.37	Vapor density[e] (g/L)	
		at 20°C	0.015
		at 178°C	3.7

[a]CRC (1990). [b]Perry (1973). [c]Lange (1985). [d]International Critical Tables (1933). [e]Braddock et al (1986). [f]FMA (1985). [g]NBS (1977). [h]Massaldi and King (1973). [i]Yaws (1995).

bacteria. For these reasons, conditions of d-limonene biodegradability in the environment need to be established.

12.3.2.1 *Health*   There is concern about human exposure and potential hazard at the manufacturing source, in transit, or at the use location, as well as environmental harm from d-limonene. General biological data of d-limonene is described as GRAS by the FDA (21 CFR § 182.60) for use in human foods. For rats, mice, and rabbits, it has low toxicity [acute oral and dermal lethal dose (LD_{50}) = 5 g/kg]. It irritates human skin on contact, but any dermatitis may be due to contaminants or oxidation products. The Environmental Protection Agency (EPA) has granted a residue tolerance exemption for d-limonene used as an inert ingredient (e.g., as a solvent or fragrance) in pesticide formulations applied pre- or postharvest to crops (Federal Register, 1994).

A two-year study of toxic effects of d-limonene reported liver effects at doses of >500 mg/kg body weight in strains of male mice and reduced survival in female rats at >600 mg/kg. In male rat kidney only, there was dose-related evidence of carcinogenic activity (NTP, 1990). However, there were no observed effects at 250 and 300 mg/kg, respectively, for the liver studies. Other studies reported that oral doses of d-limonene (100–1000 mg/kg) affected kidney weight but no observed microscopic effects in dogs (Webb et al., 1988). Based on the level of no observed effects, the EPA concluded that a tolerance was unnecessary to protect the public health for the inert ingredient application mentioned above. Also, in establishing the exemption rule, the EPA considered that the American Industrial Hygiene Association's Workplace Environmental Exposure Level (30 ppm, 8 hr) for d-limonene was also based on the no observed effect level.

There is some evidence that dietary d-limonene may afford protection from chemical tumor formation. Certain biochemical pathways in tumor tissues may be sensitive to the inhibitory action of d-limonene and certain terpene constituents of essential oils (Elson and Yu, 1994). Mammary tumor development was repressed in rats fed d-limonene during experiments to chemically induce tumors (Crowell et al., 1991). One might conclude that d-limonene is low to moderately toxic to humans, is safe for some applications (food), and requires further study for other applications because of conflicting or limited data.

12.3.2.2 *Volatile Organic Compounds*   Inefficient recovery of peel oil and d-limonene allows escape into the waste heat evaporator/dryer vent stack and contributes to release of volatile organic compounds (VOCs) into the atmosphere, a process that must be permitted by government regulation in Florida (Buff, 1997). The recent use of d-limonene as a cleaning solvent replacing chlorinated hydrocarbons has expanded its use range and precipitated a need for VOC and environmental data. Consequently, there have been experiments to gather data. One study reported the VOCs in maintenance shops using d-limonene as a degreaser. Depending on the location and type of application,

VOC concentration of d-limonene varied from 0.5 to >635 mg/m^3 of air sampled (NIOSH, 1993).

The interest in d-limonene as a VOC stems from the fact that certain chemicals deplete ozone in the stratosphere. d-Limonene will react readily with O_2, ozone, and also nitrous oxide air pollutants. Unlike chlorinated hydrocarbons, oxidation will ultimately produce CO_2, which is assimilated by plants, for example, to make sugars (and more limonene). From a manufacturing consideration, green plants (especially citrus and pine trees) produce far more environmental VOCs than any other process. There is evidence that singlet oxygen, a component in photochemical air pollution, can react with d-limonene (Rawls and Estes, 1978). Partly because of this reactivity, d-limonene has little potential to reach the stratosphere, where it can deplete ozone, contributing to global warming.

In considering d-limonene and other terpenes as VOCs, there is a problem of determining quantitative limits for setting regulations. It has been reported that the atmosphere contains 720 billion tons of carbon, 100 billion tons of which is in CO_2 pools, which move in and out of plant life. In this scheme, plant terpenes recycle in the carbon pool between 480 and 350 million tons/yr, with an average half-life of an hour (Keyser, 1991). The average atmospheric concentration is estimated to vary regionally from 0.5 to 42 ppb. For comparison, the U.S. naval stores industry recovers about 400,000 tons of terpenes annually, while annual d-limonene recovery in Florida amounts to about 10,000 tons (FCPA, 1997).

The EPA air pollution rules (Clean Air Act of 1990) intend to reduce toxic airborne chemicals and VOCs. The State of Florida, Department of Environmental Protection, regulates air pollution from citrus processing plants under these rules (Florida Administrative Code. Ch. 17-213). Citrus d-limonene is not on the EPA list of hazardous air pollutants but is considered a VOC. In Florida, which is a marginal or moderate ozone nonattainment area, Ch. 17-213 applies if a citrus processing plant emits 100 ton/yr of VOCs. Direct experimental data is needed to establish actual concentrations in citrus processing plant waste streams. However, the d-limonene content of citrus fruit has been published (see Table 11.1). Valencia oranges contain the highest amount of about 0.5 lb/box. If 75% of this amount is recovered as by-products, the unrecovered remainder could amount to 625 tons of environmental VOCs for 10 million boxes of Valencias/yr.

Recycling of peel dryer exhaust gases to the furnace has been shown to burn some d-limonene to CO_2 and water, lowering the quantity exiting the dryer as VOCs (Buff, 1997). The extent this occurs is not known and will require further study; however, the total industry VOCs may be small compared to the pulp and paper industry. There are many questions and important regulatory issues regarding the VOC status of d-limonene. For certain, the peel dryer should be studied as the major contributor in a processing plant, but this complex issue will require careful study.

12.4 UTILIZATION

Utilization of d-limonene as a chemical raw material has a long history in the flavor and terpene resin industries. More recently, other applications as a solvent and for pesticides have been documented. An excellent review (390 references) of the various chemical reactions of limonene leading to hundreds of useful chemicals has been published (Thomas and Bessière, 1989).

12.4.1 Terpene Resins

The largest use of d-limonene is as a synthetic raw material for manufacturing adhesive resins, such as the glue on labels and envelopes and adhesives for such items as disposable diapers. Terpene monomers used for resin production are pinene, dipentene from turpentine, and d-limonene. Despite the large proportion of the world's limonene used in the production of terpene resins, limonene is a minor factor, since U.S. production amounts to over 550 million lb/yr (McBride, 1990). The resins are manufactured by polymerizing the terpene in a solvent (toluene) by a Friedel–Crafts catalyst, such as aluminum chloride (ECT, 1955). The exothermic reaction is cooled (4–10°C) and when complete is quenched with water, washed to remove the catalyst, and the solvent recovered by distillation. The crude polymer is a mixture of polymeric units of from 2 to 12 monomeric limonene units. The process is complex, the solvent and feedstock flammable, and the hydrolysis products very corrosive to stainless steel (McBride, 1990).

The purity of d-limonene in the chemical reaction for resin synthesis is very important; however, the industry purifies the purchased raw material. Thus, the high-purity orange terpenes are not pure enough for this process and must be purified by the resin manufacturer. The color and odor of crude limonene is not a factor and even cold-pressed oil may be used. The resin producer also purchases product in bulk, not drums.

12.4.2 Solvents

Of the solvent uses for d-limonene, waterless hand cleaners were among the first to replace solvents such as mineral spirits. Although more expensive than mineral spirits and kerosene, d-limonene is used because of the pleasant citrus aroma and its claimed biodegradability. It probably is no more biodegradable than other hydrocarbons. Formulas for gel-type hand cleaners include saturated fatty acids, emulsifiers, d-limonene, lye, and water (Coleman, 1975). Higher d-limonene content makes these degreasing compounds more effective, as demonstrated by cleaners with 40–60% d-limonene, 10–30% surfactant, and 20–40% water (Matta, 1985). These products are effective, but some problems with their use include oxidation of the d-limonene and reaction with certain containers. In particular, polyethylene containers are a poor choice because of their solubility to d-limonene and oxygen permeability (Kutty et al., 1994).

Other solvent applications include the use of d-limonene to replace chlorinated hydrocarbons for cleaning greases and solder residues from printed circuit boards (McBride, 1990). This application is now less useful, as newer synthetic compounds are less harmful, with the same solvent power as d-limonene.

12.4.3 Flavors

Since all citrus oils contain d-limonene as the predominant volatile compound, and many other plant essential oils also have this molecule, it is a common solvent in the flavor industry. Many flavor chemicals are also synthesized from limonene using addition reactions with water, sulfur and halogens, hydrolysis, hydrogenation, boration, oxidation, and epoxide formation (Thomas and Bessière, 1989). Hydroperoxides have also been studied and isolated because of their impact on off-flavor development in products containing citrus oil flavoring agents (Clark et al., 1981; Schieberle et al., 1987). Probably the most well-known synthesis with d-limonene takes advantage of its optical activity. This reaction, conversion of d-limonene to *l*-carvone through use of nitrosyl chloride has been described (Bordenca et al., 1951).

Hydration of d-limonene produces α-terpineol. This compound has an undesirable aroma in citrus-flavored products but is a valuable flavor chemical. It is possible to produce α-terpineol and other useful value-added compounds through microbial bioconversions using d-limonene as a raw substrate (Rama Devi and Bhattacharyya, 1977). The conversion by molds of d-limonene into several products including terpineol has been reported (Bowen, 1975). Cadwallader et al. (1989) isolated a bacterium from pine sap, *Pseudomonas gladioli*, which could utilize d-limonene as a substrate to produce α-terpineol. The enzyme, α-terpineol dehydratase, isolated from this organism stereospecifically catalyzed hydration of (4R)-(+)-limonene to (4R)-(+)-α-terpineol. The enzyme was also stereoselective, since the conversion rate for hydration of (4R)-(+)-limonene was 10 times that for (4S)-(−)-limonene (Cadwallader and Braddock, 1992). Such bioconversions to utilize natural flavors and aromas from citrus raw materials have many future possibilities (Braddock and Cadwallader, 1995).

12.4.4 Other Uses

The regulatory history of d-limonene includes registration as a pesticide in 1958. Pesticide products containing d-limonene are used for flea and tick control on pets, as an insecticide spray, an outdoor dog and cat repellant, a fly repellant tablecloth, and as a mosquito larvicide as an insect repellant for use on humans. Formulations include ready-to-use solutions, emulsifiable concentrates, and granular and impregnated material, which is applied by hand both indoors and outdoors. Use limitations include label prohibition against use on weanling kittens and a caution against use of undiluted product (EPA, 1998).

Several amino alcohols have been synthesized from d-limonene (Newhall, 1959), which subsequently showed promise as sanitizing agents or mold antagonists in food processing plants (Patrick and Newhall, 1960). The use of d-limonene as an insecticide has been proposed for cockroaches and other insect pests (Dotolo, 1983). Also, it has been reported to kill fire ants and house flies based on a noncontrolled experiment where a limonene hand cleaner was poured on a fire ant mound (Sheppard, 1983). Because of the solvent potency of terpenes, it is not surprising that direct contact would kill insects. However, the questions remain as to how fruit flies, beetles, and other insects are able to survive in the presence of these chemicals in fruit fallen to the ground under a citrus tree.

REFERENCES

AOAC (1995). Recoverable oil, Method 968.20.37.1.56, *AOAC Official Methods of Analysis*, 16th ed., Arlington, VA.

Bordenca, C., Allison, R. K., and Dirstine, P. H. (1951). *l*-Carvone from d-limonene, *Ind. Eng. Chem.* **43**(5):1196–1198.

Bowen, E. R. (1975). Potential by-products from the microbial transformation of d-limonene, *Proc. FL State Hort. Soc.* **88**:305–308.

Braddock, R. J. (1982). Ultrafiltration and reverse osmosis recovery of limonene from citrus processing waste streams, *J. Food Sci.* **47**(3):946–948.

Braddock, R. J. and Adams, J. P. (1984). Recovery of citrus oils by ultrafiltration and reverse osmosis, *Food Technol.* **38**(12):109–111.

Braddock, R. J. and Cadwallader, K. R. (1995). "Bioconversion of citrus d-limonene." In *Fruit Flavors: Biogenesis, Characterization and Authentication*, R. L. Rouseff and M. M. Leahy, Eds., ACS Symp. Ser. No. 596, American Chemical Society, Washington, D.C., Chap. 13, pp. 142–148.

Braddock, R. J., Temelli, F., and Cadwallader, K. R. (1986). Citrus essential oils—a dossier for material safety data sheets, *Food Technol.* **40**(11):114–116.

Buff, D. A. (1997). VOC emissions from citrus processing plants, *Trans. Citrus Eng. Conf.* **43**:65–79.

Cadwallader, K. R. and Braddock, R. J. (1992). "Enzymatic hydration of (4R)-(+)-limonene to (4R)-(+)-α-terpineol." *Food Science and Human Nutrition.* G. Charalambous, Ed., Elsevier, Amsterdam, pp. 571–584.

Cadwallader, K. R., Braddock, R. J., Parish, M. E., and Higgins, D. P. (1989). Bioconversion of (+)-limonene by *Pseudomonas gladioli*, *J. Food Sci.* **54**(5):1241–1245.

Clark, B. C., Jr., Jones, B. B., and Iacobucci, G. A. (1981). Characterization of the hydroperoxides derived from singlet oxygen oxidation of (+)-limonene. *Tetrahedron* **37**(Suppl. 1):405–409.

Coleman, R. L. (1975). d-Limonene as a degreasing agent, *Citrus Ind. Mag.* **56**(11): 23–25.

CRC (1990). *Handbook of Chemistry & Physics*, 71st ed., CRC, Boca Raton, FL.

Crowell, P. L., Chang, R. R., Ren, Z., Elson, C. E., and Gould, M. N. (1991). Selective inhibition of isoprenylation of 21–26 kDa proteins by the anticarcinogen d-limonene and its metabolites, *J. Biol. Chem.* **266**(26):17679–17685.

Dotolo, V. (1983). Pesticides containing d-limonene, U.S. Pat. 4,379,168.

ECT (1955). "Terpene resins." In *Encyclopedia of Chemical Technology*, Vol 14, Interscience Encyclopedia, New York.

Elson, C. E. and Yu, S. G. (1994). The chemoprevention of cancer by mevalonate-derived constituents of fruits and vegetables, *J. Nutr.* **124**(5):607–614.

EPA (1998). www.epa.gov/docs/REDs/3083.

FCPA (1997). Statistical Summary, 1996–1997 Season, FL Citrus Proc. Assn, Winter Haven, FL.

Federal Register (1994). d-Limonene. Tolerance exemption, 40 CFR Part 180, *Fed. Reg.*, Vol. 59, No. 89, May 10, 24055.

FMA (1985). Flash points, closed cup, aromatic chemicals and isolates, E.O.A. No. 1-V, Fragrance Materials Association of the U.S., Washington, D.C., p. 8.

Gerow, G. P. (1974). Economics of d-limonene recovery, *Trans. Citrus Eng. Conf.* **20**: 61–66.

Graumlich, T. R. (1983). Potential fermentation products from citrus processing wastes, *Food Technol.* **37**(12):94–97.

International Critical Tables (1933). *International critical tables of numerical data, physics, chemistry and technology.* Vol. 4: p. 460; Vol. 5: p. 163, McGraw-Hill, Inc., New York, NY.

Keyser, G. E. (1991). Terpene solvents from forest products, *Naval Stores Rev. Jan/Feb*: 10–12.

Kruger, A. J., Corkum, M. L., Carlson, S. G., and Kimball, D. H. (1993). Process for purifying d-limonene, U.S. Pat. 5,220,105.

Kutty, V., Braddock, R. J., and Sadler, G. D. (1994). Oxidation of d-limonene in presence of low density polyethylene, *J. Food Sci.* **59**(2):402–405.

Lange, N. A. (1985). *Lange's Handbook of Chemistry*, 13th ed., J. A. Dean, Ed., McGraw-Hill, New York.

Massaldi, H. A. and King, C. J. (1973). Simple technique to determine solubilities of sparingly soluble organics: Solubility and activity coefficients of d-limonene, *n*-butylbenzene and *n*-hexylacetate in water and sucrose solutions, *J. Chem. Eng. Data* **18**(4):393–397.

Matta, G. B. (1985). d-Limonene based aqueous cleaning compositions, U.S. Pat. 4,511,488.

McBride, J. J. (1990). Limonene—a versatile chemical, *Trans. Citrus Eng. Conf.* **36**: 72–80.

NBS (1977). Table of dielectric constants of pure liquid, NTIS, National Bureau of Standards, Washington, D.C.

Newhall, W. F. (1959). Derivatives of (+)-limonene II. 2-Amino-1-*p*-menthanols, *J. Org. Chem.* **24**:1673–1676.

NFPA (1975). *Fire Protection Guide on Hazardous Materials*, 6th ed., National Fire Protection Association, Boston.

NIOSH (1993). Hazard evaluation and technical assistance report no. 92-0101-2341, U.S. Dept. Heath & Human Services, Public Health Service, Region IV, Atlanta, GA, August.

NTP (1990). Toxicology and carcinogenesis studies of d-limonene, U.S. Dept. Health & Human Services, Public Health Service, NIH Pub. No. 90-2802, NTP Technical Report No. 347, Research Triangle Park, NC.

Odio, C. E. (1993). Tank reaction system for citrus peel, *Trans. Citrus Eng. Conf.* **39**: 1–14.

Patrick, R. and Newhall, W. F. (1960). Fungicidal activity of some new amino alcohols synthesized from citrus (+)-limonene, *J. Agr. Food Chem.* **8**(5):397–399.

Perry, R. H. (1973). *Chemical Engineers' Handbook*, 5th ed., McGraw-Hill, New York, NY.

Rama Devi, J. and Bhattacharyya, P. K. (1977). Microbiological transformations of terpenes: Part XXIV. Pathways of degradation of geraniol, nerol and limonene by *Pseudomonas incognita* (linalool strain), *Indian J. Biochem. Biophys.* **14**:359–363.

Rawls, H. L. and Estes, F. L. (1978). A preliminary investigation of the use of limonene for detecting singlet oxygen as a component of polluted air, *Photochem. Photobiol.* **28**:465–468.

Schieberle, P., Maier, W., Firi, J., and Grosch, W. (1987). HRGC separation of hydroperoxides formed during the photosensitized oxidation of (R)-(+)-limonene, *J. High Resolution Chromatog. Chromatog. Commun.* **10**(11):588–593.

Schulz, H. E. (1972). d-Limonene recovery in the Florida citrus industry, *Trans. Citrus Eng. Conf.* **39**:1–6.

Scott, W. C. and Veldhuis, M. K. (1966). Rapid estimation of recoverable oil in citrus juices by bromate titration, *J. A.O.A.C.* **49**(3):628–633.

Sheppard, D. C. (1983). Toxicity of citrus peel liquids to the house fly and red imported fire ant, *J. Agric. Entomol.* **1**(2):95–100.

Thomas, A. F. and Bessière, Y. (1989). Limonene, *Natural Products Rep.* **6**(3):291–309.

Webb, D. R., Hysell, D. K., and Alden, C. L. (1988). Assessment of the chronic oral toxicity of d-limonene in dogs, Abstr. 356, *Toxicologist* **8**(1):90.

Yaws, C. L. (1995). *d-Limonene. Handbook of Viscosity*, Vol. 3, Gulf Publishing Co., Houston, TX, p. 197.

CHAPTER 13

PECTIN

Citrus flavedo, albedo, membranes, juice vesicles, and core contain significant quantities of pectin, in the form called protopectin. There is significant variation in pectin quantity and quality within these tissues as well as between the varieties. Generally, the quantity and quality of pectin decreases beginning with lime > lemon > grapefruit > orange > mandarin-type fruits. Within the fruit itself, the albedo is the source of the highest quantity and quality of pectin and the juice vesicles the lowest. The world market for pectin is limited (pectin is also made from apple pomace), which means that only moderate tonnages of the total amount of citrus peel can be used as a pectin feedstock. The extraction process also is waste intensive, which limits production to specific locales.

13.1 MANUFACTURE

13.1.1 Extraction

Citrus fruit peels contain approximately 30% of the dry matter as pectin, which is usually extracted from either the wet or dried peel. If wet peel is used, the pectin recovery must be performed near the citrus processing plant as the peel rapidly ferments and spoils. If dried peel is used, it must be carefully washed free of soluble sugars to prevent browning during drying. The general details of pectin extraction and recovery from citrus fruit have been previously described (Snyder, 1970; Sinclair, 1972). A complete review of its occurrence in citrus fruits, functionality in juice processing, and methods for extraction and recovery as a by-product has been published (Rouse, 1977). A flow diagram of the procedure is presented in Figure 13.1. The pectin is extracted from the

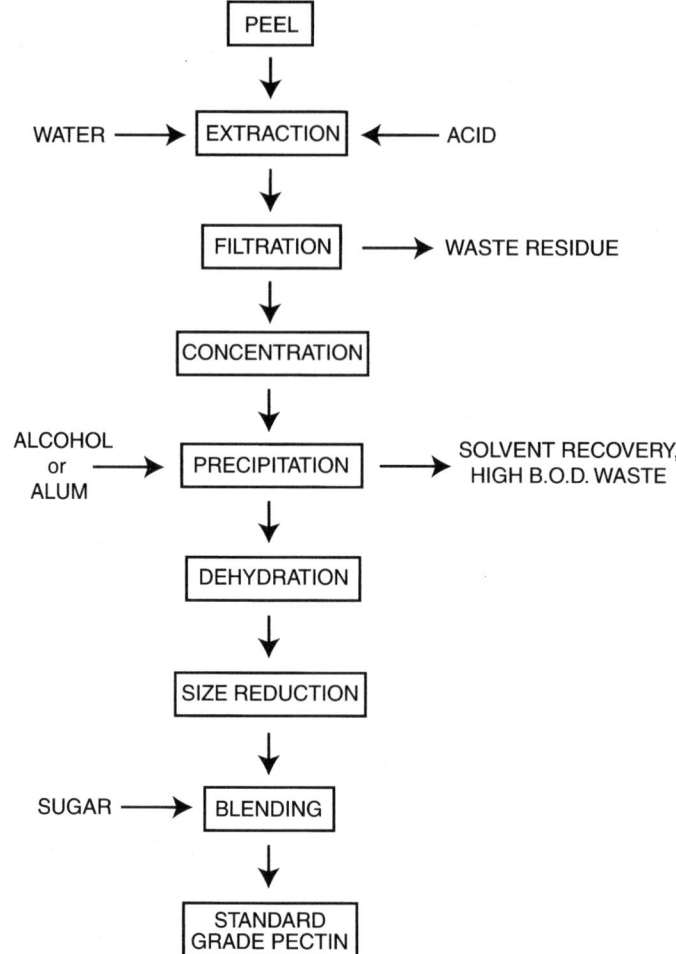

Figure 13.1   Generalized flow procedure for citrus pectin manufacture.

dried peel (or wet) with a mixture of water and a dilute mineral acid (nitric, sulfuric, or hydrochloric) for several hours at controlled temperatures between 50 and 100°C. The process is controlled between pH 2 and 3 by dilution to solubilize the protopectin. After solubilization, the pectin solution is filtered or centrifuged to remove insoluble peel particles and concentrated to about 3–5% by vacuum evaporation.

13.1.2   Recovery

Since most pectin is sold as a powdered product, the pectin must be precipitated from solution, by either alcohol or aluminum salts. Isopropanol or ethanol added to the extent of about 60% in the pectin solution precipitates the pectin, which is then washed several times with increasing concentrations of acidic alcohol. In the so-called alum process, aluminum chloride or sulfate precipitates the pectin as aluminum pectinate, which is also washed with acidic alcohol for purification. After the final wash, the pectin is pressed, dried under vacuum or with circulating air/nitrogen to a moisture content of 6–10%, and ground to a powder (Sinclair, 1972). The product is blended with dextrose or sucrose for standardization (Fig. 13.1). This process is complicated and takes several days, in addition to the expense of recovering the solvents in a distillation column, and disposing of the extracted residue, high BOD dilute sugars, and waste acid.

13.1.3 Pectin Pomace

It is not always possible to have the pectin factory located near a citrus processing plant as a seasonal source of raw peel. Thus, pectin pomace, a dried peel product for pectin feedstock is produced, mostly from lemon and lime peel, because of the higher yields and quality. As performed today, the process involves two-stage countercurrent water washing and pressing to remove the soluble peel sugars and acid, before drying in the same kiln-type dryers used for dried pulp manufacture. An earlier process for grapefruit peel pectin used hot water to inactivate the pectinesterase and solubilize the naringin, which could be filtered and precipitated as a by-product (Poore, 1934). Commercial production of a low-jelly-grade pectin pomace from grapefruit peel residue was also described (Pulley et al., 1944).

Because no calcium oxide is used as a pressing aid as in dried pulp manufacture, the pectin binds water as the sugars are leached, resulting in press cake moisture contents in the 85–90% range. This means that the wet press cake will stick to hot dryer surfaces near the furnace and burn, resulting in undesirable black pieces in the dried pomace. For this reason, dryer temperatures and capacities must be lowered to almost half that used in drying cattle feed, making the product cost much higher. The drying process also results in 25% yield loss of pectin after the extraction process than if fresh, wet peel was used (Crandall et al., 1978). However, the advantage of storing and shipping the dry pectin feedstock must be considered.

13.1.4 Standardization

13.1.4.1 Yield After extraction, precipitation, and purification, the dried pectin yield is about 25–30% of the dry weight of the peel. Several values must be considered though, as the mass of pectin is not the only important variable. The yield of pectin varies between varieties as does the quality, which determines its properties for gelation. Thus, procedures to standardize the pectin by diluting the dried product with sugar have been established, which also allows yield normalization. In general, there are about 0.9–1.4 kg (2–3 lb) of 150-grade pectin per box of citrus. The fresh weight pectin content of the edible portions and some parts of many fruits and vegetables, including citrus, has been compiled in a critical review (Baker, 1997).

13.1.4.2 Jelly Grade   Jelly grade is the grams of sugar, which can be made into a standard gel by one gram of pectin. Jelly grade of pectin is determined by a standardized procedure, which is based on the ratio between the total soluble solids and the grams of pectin used and a relationship to the amount of sag of a jelly released by inverting a glass of standard 65 °Brix jelly by a device called a Ridgelimeter (IFT, 1959). The procedure assumes 650 g of total soluble solids (including the pectin) in 1000 g of jelly. Thus, 1000 g of jelly made with 150-jelly-grade pectin contains 4.33 g of pectin (650/150). The jelly grade of pure extracted citrus pectin may range from 150 to 300, depending on many process and fruit variables; therefore, the dried, purified pectin is commonly diluted with sugar to a standard 150 grade.

13.1.4.3 Jelly Units   If one multiplies the yield of pectin after the extraction-purification process by the jelly grade, the result is termed the jelly units (e.g., 0.25×220 grade lemon pectin = 55 jelly units). This value is helpful when considering the amount and quality of pectin from an extraction process to make a standard jelly. For example, the same amount of process chemicals and cost for a 25% yield of 180-grade pectin from orange will only yield 45 jelly units.

13.2 COMPOSITION

13.2.1 Purity

Pectin is a polymer composed of galacturonic acid residues, a percentage of which are esterified with methanol. Pectin may be analyzed by a colorimetric method to determine the anhydrogalacturonic acid (AGA) content. The most common of these methods, the Carbazole test, is based on the quantitative formation of pink color by hexuronic acids with carbazole in an acid solution (Dische, 1947). By this analysis, the purest citrus pectins contain about 85–90% AGA. The remainder of the pectin consists of sugars (rhamnose, xylose, arabinose, and galactose) and methyl esters of the galacturonic acid.

13.2.2 Degree of Methoxylation

A percentage of the galacturonic acid residues of the pectin molecule are esterified with methanol, measured by the degree of methoxylation (DM). The highest DM for citrus pectin is about 70–80%, and this pectin would be considered "high-methoxyl" pectin. This pectin would be considered "rapid-set" pectin if used to make a jelly. The higher the DM, the faster the gel will set (harden). Gels made with high-methoxyl pectin set at pH = 2.8–3.4. Some high-methoxyl pectins with lower DM (60–65%) are termed "slow-set" pectins. The pH range of 2.8–3.2 is optimum for this pectin to gel. If enough of the methoxyl groups are hydrolyzed from the pectin molecule by acid (or pectinesterase), low DM, or "low-methoxyl" (LM) pectin results. This is usually in the range DM < 50. Although, because more carboxyl groups are present, addition of divalent metal ion salts (calcium) allow gel formation at wide pH range (2.8–6.5) and application to foods with higher pH.

The polymer is termed pectic acid if none of the residues are esterified. Pectic acid and the LM pectins may gel in the presence of calcium, which can have a bitter taste at higher concentrations in some foods. If lower calcium concentrations are required, low DM pectin may be amidated by de-esterification with ammonia. Amidated pectins gel at higher pH with a lower calcium requirement.

13.3   UTILIZATION

13.3.1   Jelly Products

The major use of pectin is for the manufacture of jelly, jams, preserves, marmalade, and fruit butters. Clear jelly can be made with slow-set pectin, which allows long enough setting times to let small air bubbles leave the jelly. Rapid-set pectins are used in the preparation of jams, preserves, and marmalade to prevent the fruit pieces from floating. Conventional jellies are made to 65 °Brix with sucrose as the sugar, but the demand for dietetic, lower sugar products may use other sweeteners. These products require LM pectin.

13.3.2   Dairy Gels and Desserts

High- and low-methoxyl pectins are used to stabilize fermented dairy products, such as yogurt. Use of pectin allows stabilization of blended fruit juice/milk products because the milk protein, casein, precipitates in the presence of the fruit acid (Nelson, 1979). The mechanism of stabilization involves the positive charge of casein below its isoelectric point (pH 4.6) being stabilized by the net negative charge of pectin, forming a casein–pectin complex. The presence of calcium favors gelation of these products. The LM pectins are used in fruit and vegetable gels served as salads as well as instant puddings made with milk, starch, and other gels because of the wide pH and solids conditions of gelation.

13.3.3   Salad Dressing

Although carageenan and xanthan gums are used in mayonnaise and salad dressing, pectin finds application in pourable products. Its primary use is as a thickening agent and emulsifier for the fat component of these products.

13.3.4   Pharmaceutical Products

Pectin and some of its chemical derivatives, such as metal pectinates, have clinical applications for treatment of gastrointestinal problems (Rouse, 1977). Pectin has also been considered in the class of dietary fibers, which reportedly can lower cholesterol, adjust serum glucose levels, and affect other digestive processes (Baker, 1994).

13.3.5   Other Applications

Mechanically and chemically modified pectins have many uses, including application as a fat substitute. The product, Slendid, is one of these specialty pectins used in low-calorie foods. It functions mainly as a texturizing agent to mimic the mouth-feel and properties that lipids impart to foods and can be substituted for all of a food's fat (Hoefler et al., 1993). Pectic acid and shorter chain polygalacturonic acids have also been proposed as clarification agents to precipitate the cloudiness of citrus and other fruit juices (Baker, 1976). The LM pectins also find use in fillings in bakery products, gel-type confections, and candies and as a coating or glaze for these products. Since many food applications of LM pectin require calcium, to avoid problems, the calcium is usually added to the acid during the blending process.

REFERENCES

Baker, R. A. (1976). Clarification of citrus juices with polygalacturonic acid, *J. Food Sci.* **41**:1198–1200.

Baker, R. A. (1994). Potential dietary benefits of citrus pectin and fiber, *Food Technol.* **48**(11):133–139.

Baker, R. A. (1997). Reassessment of some fruit and vegetable pectin levels, *J. Food Sci.* **62**(2):225–229.

Crandall, P. G., Braddock, R. J., and Rouse, A. H. (1978). Effect of drying on pectin made from lime and lemon pomace, *J. Food Sci.* **43**:1680–1682.

Dische, Z. (1947). A new specific color reaction of hexuronic acids, *J. Biol. Chem.* **167**: 189–198.

Hoefler, A. C., Sleap, J. A., and Trudso, J. E. (1993). Fat substitute, U.S. Pat. 5,192,575.

IFT (1959). Pectin standardization, final report of the IFT committee, *Food Technol.* **13**:496–500.

Nelson, F. F. (1979). Newer applications for pectin, *Food Prod. Develop.* **13**(2):36, 38.

Poore, H. D. (1934). Recovery of naringin and pectin from grapefruit residue, *Ind. Eng. Chem.* **26**(6):637–639.

Pulley, G. N., Moore, E. L., and Atkins, C. D. (1944). Grapefruit cannery waste yields crude citrus pectin, *Food Ind.* **16**(4):285–287, 327–328.

Rouse, A. H. (1977). "Pectin: Distribution, significance." In *Citrus Science and Technology*, Vol. 1, S. Nagy, P. E. Shaw, and M. K. Veldhuis, Eds., AVI, Westport, CT, Chap. 4, pp. 110–207.

Sinclair, W. B. (1972). "Grapefruit by-products. Pectin." In *The Grapefruit. Its Composition, Physiology and Products*, University of California Press, Riverside, CA, Chap. 6, pp. 524–530.

Snyder, R. P. (1970). Citrus pectin and dried pectin peel, *Trans. Citrus Eng. Conf.* **16**: 79–89.

CHAPTER 14

BIOCONVERSION PRODUCTS

Citrus processing waste and by-products streams are potential sources of substrates for manufacture of fermentation and bioconversion products. The large majority of the solid peel waste is manufactured into dried pulp cattle feed, with molasses added. However, a portion of the molasses may be fermented to make alcohol. Other streams, such as the oil mill effluent, d-limonene, and aqueous-phase essence also offer bioconversion potential.

14.1 ALCOHOL

Citrus molasses has been used as a substrate for the bioconversion of peel sugars to ethanol for over 50 years. This ethanol is produced in a traditional batch yeast fermentation from the molasses and is primarily used as a neutral spirits for manufacture of alcoholic beverages.

14.1.1 Raw Material

While traditional 72 °Brix citrus molasses is the preferred substrate for alcohol fermentation, use of 20–50 °Brix waste heat evaporator molasses is common in the industry. Very few traditional 72 °Brix molasses evaporators are still in use. The use of lower °Brix waste heat molasses is not as satisfactory as the high °Brix product because the peel oil remaining in the former has an inhibitory effect on yeast growth. The waste heat molasses may have from 0.2 to 0.4% peel oil; while high temperature and steam stripping during 72 °Brix molasses production reduces the oil content to near 0.001%. Unsanitary handling of the press liquor in large tanks at citrus feed mills also allows fermentation of the sugars, producing alcohol and CO_2, which is lost during molasses concentration in the waste heat evaporator. This fermentation uses the most fermentable sugars, which then are unavailable to the distiller purchasing the molasses. Alcohol in the waste streams and condensate from the evaporator also presents potential problems for the citrus processor, who might lose d-limonene or have waste treatment problems.

Distillers or citrus processors with alcohol plants using waste heat molasses usually have evaporative capacity to concentrate this molasses to 72 °Brix. This concentration strips the remainder of the peel oil from the product and also allows long-term storage as a fermentation feedstock for use during times when waste heat citrus molasses is unavailable. With some operational knowledge related to scale formation and manufacturer modifications to adjust for higher viscosities, the traditional TASTE juice evaporators may be used for this purpose.

14.1.2   Manufacture

The molasses is diluted with water to near 25 °Brix in large 50,000-gal (190,000-L) fermentation tanks, the yeast culture added, and the fermentation allowed to proceed for several days. The process flow diagram of Figure 14.1 is typical of this fermentation, which distills the fermented molasses in a beer

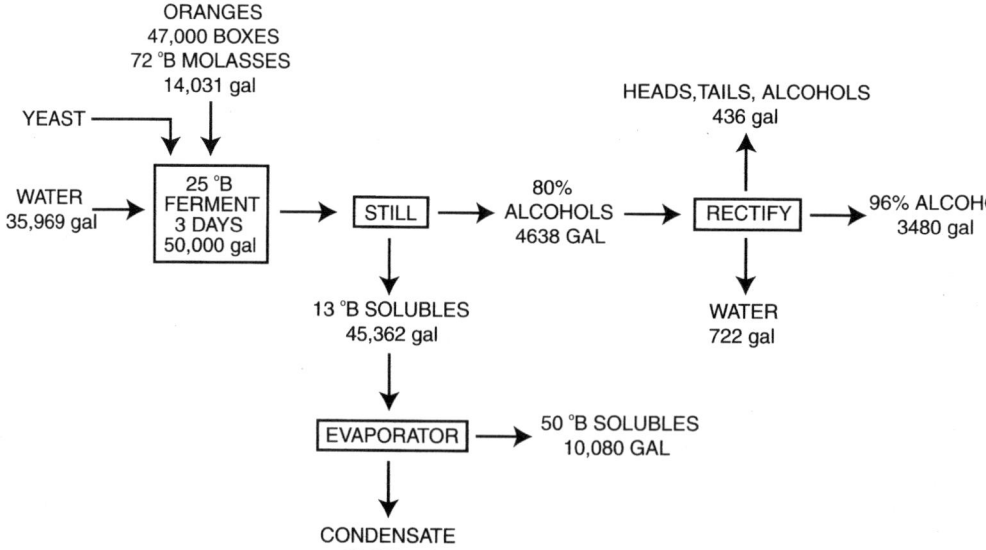

Figure 14.1   Commercial fermentation process for alcohol from citrus molasses.

TABLE 14.1 Analysis of Citrus Pellets with 50 °Brix Molasses (M) or Distiller's Residuum (R) Added at 0.25 Parts M or R per One Part Press Cake (wt/wt)

Property	M	R
Moisture (%)	10	10
NFE[a] (%)	64.0	60.3
Crude fiber (%)	10.8	10.9
Protein (%)	5.6	6.5
Fat (%)	2.2	2.0
Ash (%)	7.3	10.3

[a]NFE, nitrogen-free extract.

still to recover approximately 80% alcohol. The alcohol from the still is rectified in a distillation column that removes fusel oils (heads, tails, 5-carbon alcohols) and water, concentrating the product to 95–96% alcohol. The beer still residuals from the alcohol recovery include unfermented sugars and yeast protein, which can be concentrated to near 50 °Brix in an evaporator. This evaporator, with proper scheduling, is usually the same one used to concentrate waste heat molasses to 72 °Brix. The higher protein content of the residuum significantly increases viscosity and limits the concentration to about 50–55 °Brix.

The concentrated residuum may be sold as a liquid cattle feed supplement, sent back to the citrus plant for addition to the press cake entering the dryer, or disposed of by land spreading techniques. The 50 °Brix residuum may be successfully added to the peel in the feed mill with little difference in handling, compared to traditional waste heat molasses added at a mass rate of 0.2–0.3 parts residuum to one part press cake. Concerns that the dry pellets with added residuum would differ in composition from that with molasses are probably not verifiable by analytical data because of the small total amount of residuum product, compared to the total. Some analytical results indicate the small differences between pellets with molasses and residuum (Table 14.1).

14.1.3 Utilization

The primary use for alcohol from citrus molasses is as a neutral alcohol feedstock to make alcoholic spirits, liquors, and cordials. Because the beverage alcohol market is limited, there has been occasional interest in making citrus alcohol for use as a fuel (as for example, mixing with gasoline). However, the raw material value as a cattle feed and the cost of the alcohol manufacturing process makes this economically questionable. There is still interest in this possibility, involving research to make conversion of molasses and citrus peel

TABLE 14.2 Calculated Alcohol Yield from Orange By-products

Product	Gal Alcohol/Box[a]
Juice	0.39
Pulp wash	0.03
Essence	0.0009
Molasses	0.06
Peel (sugar + fiber)[b]	0.14

[a]Theory of 14 lb sugar/gal 100% ethanol and by-product solids yield (Chapters 3 and 7).
[b]Grohmann et al. (1996). Estimate that peel and fiber is 20% solids, 80% water, and yielded 38 g alcohol from 111 g total peel hydrolysate of sugar and fiber.

sugars and carbohydrate fibers more efficient than the traditional <50% conversion of sugar to alcohol by batch yeast fermentation. A traditional process theoretically requires about 14 lb sugar to produce a gallon of 100% ethanol. Some proposed processes allow conversion of citrus peel fibers and 5-carbon sugars, improving the alcohol yield above that from only the fermentable sugars (Grohmann et al., 1996). Some calculations of the yield of alcohol from the various citrus products are presented in Table 14.2. This table illustrates only that the two products with the highest sugar content, juice and peel, yield the most alcohol. If one applies economics to the value of the products, the traditional food uses are still favored over conversion to fuel alcohol.

14.2 WINE

Citrus juices can be converted to suitable wines but only by amelioration with sugar and water to dilute the strong citrus flavor, which oxidizes during storage of the wines (von Loesecke et al., 1936). The finished wines do not possess distinct citrus flavor and aroma, characteristic of the fruit source. Bitterness from naringin also can be detected in grapefruit wines. Out-of-grade orange juice concentrate may be fermented to make wine, which is distilled and the distillate used to make wine coolers, spirits, and brandy.

14.3 VINEGAR

Citrus vinegar made from oranges, grapefruits, and tangerines has been reported to possess an interesting and distinctive flavor (McNary and Dougherty, 1960).

Citrus juices used for vinegar production do not contain sufficient sugar to yield 4–5% acetic acid after the wine fermentation, without adding sugar to the fermentation, or concentration of the alcohol.

14.4 2,3-BUTYLENE GLYCOL

Interest in fermenting the press liquor to make 2,3-butylene glycol resulted in research to make this useful chemical feedstock (Long and Patrick, 1963). From 20 °Brix press liquor, the ideal fermentation produced about 5% glycol in 2–4 days at 30°C. This chemical has a number of industrial uses as follows: textile softeners, dye solvents, synthetic resins, and antifreeze. However, economic recovery of the product from the fermentation residual proved costly.

14.5 PROTEIN

Citrus waste streams have been studied for their potential to produce single-cell protein for both human and animal foods. A number of such studies were reviewed, with results of yeast and mold dry matter yields (protein content of 45%) between 2 and 4 kg/100 L of 6–10% solids waste streams (Graumlich, 1983). This author also reported a number of other interesting citrus-related microbial research studies. Attempts to increase the protein content of citrus peel residue have also been reported. Solid-state fermentation of extractor residues and finisher pulp by *Aspergillis niger* was capable of producing a pectinase that could break down the cellular structure of peel (Wicker et al., 1987). This process also resulted in increasing the protein content of the residue five-fold. Finisher pulp was also considered a viable substrate for production of polygalacturonase by fungal growth (Hart et al., 1991). Protein content was enriched from the basic value of 6% to near 55% (dry weight) in a batch fermentation of orange peel by *Saccharomyces* spp. (Nwabueze and Oguntimein, 1987). Solid-state fermentation is a simpler process than liquid fermentation and may be used to directly convert peel carbohydrates, without extraction, to protein, thus enriching the nutritional value (Nicolini et al., 1987).

14.6 METHANE

Potential production of methane from citrus wastes for energy uses in processing and elimination of pollution has a long history. Early studies attempted to use the press liquor in order to avoid the cost of concentration to molasses. However, the organisms in the anaerobic digestion required were quite sensitive to the peel oils in various feedstreams, down to 0.0002%, v/v (McNary et al., 1951). These authors proposed allowing yeast fermentation to use some of the solids and reduce the oil content, followed by anaerobic digestion to produce methane. Later studies revisited this subject and also decided that the peel oils were poisonous to activated sludge waste treatment of citrus wastes with the object of methane production (Lane, 1984). Processes to aerate, distill, or centrifuge waste streams to lower the oil concentration to acceptable levels make methane production uneconomical.

14.7   WASTE TREATMENT

Citrus processors use various combinations of waste treatment, including activated sludge. Plants using activated sludge treatment usually route the waste stream to an aeration process, which initiates the treatment and regulates day-to-day loading consistency. A secondary aeration allows significant culture growth and sludge production. The stream is then clarified by settling and removal of the sludge by sand filtration, where a portion is recycled back to the system. Dried sludge is high in protein (20–45%) and contains a number of vitamins (especially B vitamins) produced by the organisms in the process (Kesterson and Braddock, 1976).

Liquid wastes may be sent to a lagoon system designed to remove soluble and insoluble materials. Treatment consists of a series of aerated lagoons, settling, and polishing ponds. From the aerated lagoons, effluent may flow to an anaerobic lagoon, where algae and scum seal the surface, allowing solids and matter to sink and decay. Decay produces gas which bubbles to the surface with particles, helping to seal the surface. Excess material is removed, and the liquid effluent passes to other ponds where the aeration-anaerobic cycle may be repeated many times. Clear water from such lagoons may undergo a final treatment or in some locations be discharged. All of these processes are subject to local and governmental regulations.

Alternatively, in suitable locations, dilute streams (below 1 °Brix) from the processing plant may be collected, passed over an inclined screen to remove particles, and disposed of by land spreading or spray irrigation (Hong, 1977). Spray fields may be from 10 to 100 acres and have capacity to percolate millions of gallons of liquid/day. The field is divided into sections, which are used on a rotation basis to prevent soil plugging from fermenting solids. If any of the waste liquids are used for crop irrigation, careful monitoring and computer control of the nutrient content is essential to prevent damage to the crop. Sodium from lye cleaning of process equipment may concentrate in the soil and is a particular problem. Such monitoring may even allow adjustment of the fertilizer formula for the intended crop, if certain required nutrients are of adequate concentration in the processing waste stream. In the future, citrus processing waste treatment will environmentally integrate processing with fruit production and minimize, or recycle, all of the effluent.

14.8 OTHER PRODUCTS

14.8.1 Peel Oil Emulsion

High biological oxygen demand (BOD) levels of citrus wastes offer many possibilities for microbial conversions. The peel oil emulsion stream from juice extractors has chemical oxygen demands (CODs) and BODs of 30,000 and 11,000 mg/L, respectively (Crandall and Kesterson, 1980). Other experimental studies considered the peel oil recovery streams as sources of fermentation substrates for a variety of potential bioconversion products. Centrifuge aqueous discharges contained 3.8 °Brix, 2.6% sugars, 0.28% oil, and a BOD of 22,600 mg/L (Parish et al., 1986a). After d-limonene was removed by distillation, this centrifuge aqueous discharge was fermented by several organisms in a plan to produce alcohol (3%), lipid (26% of cell dry mass), or crude extracellular polysaccharides (0.3 g/100 mL effluent) (Parish et al., 1986b). Residue from the distillation of lime oil also has been reported to be a substrate for the production of vitamin B_{12} via the growth of *Propionibacteria* strains (Perez-Mendoza and Garcia-Hernandez, 1983).

14.8.2 Flavors

Many of the compounds in citrus oils and essences are chiral, with the enantiomeric ratio specific to the source of the oil. While this is useful for detecting adulteration and purity, it also has application for biosynthesis of chiral flavor molecules with unique aroma properties. Bioconversion of flavors from natural citrus oils has the specific feature of producing value-added natural flavor compounds. In the United States, it is stated (21 CFR § 101.22) that chemicals produced by enzymes or living cells may be considered natural for labeling purposes if the starting material is natural. A discussion of bioconversion of citrus d-limonene to the more valuable *d*-α-terpineol has been presented in Chapter 12.

Natural ethanol in aqueous citrus essence can be oxidized by bioconversion to natural acetaldehyde, enhancing the fresh notes of the essence aroma (Goodrich et al., 1998). This process, utilizing the methylotrophic yeast *Pichia pastoris*, is capable of bioconversion of essence ethanol, propanol, and octanol to their respective aldehydes. Biomodified essence contained 1200 mg/L acetaldehyde concentration compared with the control (600 mg/L). The proposed whole cell process also has the advantage of self-generation of expensive cofactors required in similar alcohol dehydrogenase systems (Raymond, 1984).

14.8.3 Amino Acids and Riboflavin

Other potential products reviewed include production of amino acids from molasses, pyruvic acid by yeast fermentation from peel extracts, and a microbial digestion, which produced an enzyme to solubilize protopectin, allowing less costly pectin recovery (Graumlich, 1983). Production of riboflavin from clarified, diluted citrus molasses (3.3 °Brix) has been demonstrated. By supplementation with yeast extract, the organism *Eremothecium ashbyii* was capable of producing 0.7 g riboflavin/L dilute molasses substrate over a 7-day fermentation (Gaden et al., 1954).

REFERENCES

Crandall, P. G. and Kesterson, J. W. (1980). BOD and COD determinations on citrus waste streams and component parts, *J. Food Sci.* **45**:134–137.

Gaden, E. L., Jr., Petsiavas, D. N., and Winoker, J. (1954). Microbiological production of riboflavin and citric acid from citrus molasses, *J. Agr. Food Chem.* **2**:632–638.

Goodrich, R. M., Braddock, R. J., Parish, M. E., and Sims, C. A. (1998). Bioconversion of citrus aroma compounds by *Pichia pastoris, J. Food Sci.* **63**(3):445–449.

Graumlich, T. R. (1983). Potential fermentation products from citrus processing wastes, *Food Technol.* **37**(12):94–97.

Grohmann, K., Cameron, R. G., and Buslig, B. S. (1996). Fermentation of orange peel hydrolysates by ethanologenic *Escherichia coli, Appl. Biochem. Biotechnol.* **57/58**: 383–388.

Hart, H. E., Parish, M. E., Burns, J. K., and Wicker, L. (1991). Orange finisher pulp as substrate for polygalacturonase production by *Rhizopus oryzae, J. Food Sci.* **56**(2): 480–483.

Hong, T. (1977). Wastewater treatment and recycling in the citrus industry, *Proc. Intl. Soc. Citriculture* **3**:768–773.

Kesterson, J. W. and Braddock, R. J. (1976). By-products and specialty products of Florida citrus, FL Agr. Exp. Sta. Tech. Bull. No. 784, University of Florida, Gainesville, FL.

Lane, A. G. (1984). Anaerobic digestion of orange peel, *Food Technol. Australia* **36**(3): 125–127.

Long, S. K. and Patrick, R. (1963). The present status of the 2,3-butylene glycol fermentation, *Adv. Appl. Microbiol.* **5**:135–155.

McNary, R. R. and Dougherty, M. H. (1960). Citrus vinegar, FL Agr. Exp. Sta. Bull. No. 622, University of Florida, IFAS, Gainesville, FL.

McNary, R. R., Wolford, R. W., and Patton, V. D. (1951). Experimental treatment of citrus waste water, *Food Technol.* **5**(8):319–323.

Nicolini, L., von Hunolstein, C., and Carilli, A. (1987). Solid state fermentation of orange peel and grape stalks by *Pleurotus ostreatus, Agrocybe aegerita* and *Armillariella mellea, Appl. Microbiol. Biotechnol.* **26**:95–98.

Nwabueze, T. U. and Oguntimein, G. B. (1987). Sweet orange residue as a substrate for single cell protein production, *Biolog. Wastes* **20**:71–75.

Parish, M. E., Braddock, R. J., and Graumlich, T. R. (1986a). Chemical and microbial characterization of citrus oil-mill effluent, *J. Food Sci.* **51**(2):431–433.

Parish, M. E., Braddock, R. J., and Graumlich, T. R. (1986b). Potential microbial utilization of citrus oil-mill effluent, *J. Food Sci.* **51**(3):839–840.

Perez-Mendoza, J. L. and Garcia-Hernandez, F. (1983). Fermentation of a waste product from the industrial processing of the lime (*Citrus aurantifolia* Swingle) for vitamin B_{12} production, *Biotechnol. Let.* **5**(4):259–264.

Raymond, W. R. (1984). Process for the generation of acetaldehyde from ethanol, U.S. Pat. 4,481,292.

von Loesecke, H. W., Mottern, H. H., and Pulley, G. N. (1936). Wines, brandies and cordials from citrus fruits, *Ind. Eng. Chem.* **28**:1224–1229.

Wicker, L., Parish, M. E., and Braddock, R. J. (1987). Nonconventional treatment of citrus juice extractor residue, *Proc. FL State Hort. Soc.* **100**:44–46.

CHAPTER 15

FLAVONOIDS AND LIMONOIDS

Particularly high concentrations of the flavanone glycosides, hesperidin, and naringin and lesser quantities of many other flavonoids are found in citrus fruit. In mature fruit, flavonoid concentration is highest in the tissue (albedo, 30–50% of total; rag, core, and pulp, 30–50%; flavedo, 10–20%) and lowest in the juice (1–5%). The properties and chemistry of citrus flavonoids have been studied extensively and a number of reviews published (Horowitz and Gentili, 1977). Limonin is the principle limonoid in citrus and is in highest concentration in the seeds and similar to the flavonoids, in the tissue, with lower amounts in the juice.

15.1 HESPERIDIN

Hesperidin occurs in sweet and sour oranges, lemons, limes, mandarins, and in leaves, twigs, bark, and blossoms. The biological purpose of hesperidin in citrus has not been identified; however, specific flavonoids may have taxonomic significance. There has been considerable attention given to the pharmacological, or medicinal, properties of hesperidin and other flavonoids since the extraction from lemons of a vitaminlike substance, vitamin P, now referred to as "bioflavonoids" (Szent-Gyorgyi, 1938). A quick check of the current literature (and the Internet) will indicate that interest in the medicinal properties of hesperidin and citrus bioflavonoids has not waned.

15.1.1 Recovery

The hesperidin content of the component parts of the major citrus fruits is listed in Table 15.1. These values represent about 1.0 g of hesperidin per orange

TABLE 15.1   Average Flavonoid[a] Content of Citrus Fruit Component Parts

Fruit	Lb Flavonoid/Box Fruit				
	Albedo[b]	Flavedo	Rag & Pulp	Juice	Total
Early-mid	0.20	0.08	0.16	0.01	0.45
Valencia	0.11	0.07	0.17	0.01	0.37
Dancy	—	0.10	0.37	0.02	0.49
Temple	0.19	0.06	0.14	0.02	0.42
Lemon	0.12	0.10	0.16	0.02	0.40
Marsh	0.17	0.02	0.11	0.01	0.31
Ruby red	0.22	0.02	0.11	0.01	0.37

[a]Hesperidin in oranges, tangerines, and lemons; naringin in grapefruits.
[b]The mass distribution and percent dry weight of the fruit components are presented in Tables 3.1–3.4.

or 9–10 lb/ton of fruit from the residue of peel, rag, and pulp. The percentage of hesperidin in the fruit decreases dramatically as the fruit increases in size and matures. Fruit of 0.5–1 inch diameter contain almost 10% of the dry weight as hesperidin, decreasing to <0.5% at full maturity (Hendrickson and Kesterson, 1964). The traditional method for hesperidin analyses, the Davis test, is based on measuring the intensity at 420 nm of the yellow color of flavonoids in basic solutions (Davis, 1947). Modern determinations usually involve HPLC of extracts (Manthey and Grohmann, 1996).

Practical methods for commercial recovery of hesperidin involve extraction of the peel residue at pH 11.0 with calcium hydroxide (Higby, 1947; Hendrickson and Kesterson, 1954). Use of lime, rather than NaOH, is necessary to allow release of bound water and prevent solubilization and hydrolysis of pectin, avoiding large viscosity increases and slurry handling problems. After extraction, the alkaline slurry is clarified by pressing and filtration, and the solution is adjusted to pH 4.0–4.5. This solution is held for 24–48 hr, allowing hesperidin to crystallize, after which it is recovered by filtration. The product, crude hesperidin (about 40%), is dried and can be further purified. Residual hesperidin in the filtrate may be recovered by concentration of the filtrate and recrystallization, since losses could be as high as 40–50% of the total in the residue. Purification of crude hesperidin is accomplished by dissolution in 0.2 N NaOH containing 50% isopropanol to yield a 2% solution, filtration, and adjustment to pH 8.5, allowing crystallization for 48 hr. The purified crystals are recovered by filtration and dried. This process may be repeated, if required, to produce >95% purity hesperidin. An alternate method involves dissolving 10% crude hesperidin in formamide and recrystallization by dilution with water (Pritchett and Merchant, 1946).

Hesperidin has also been commercially recovered from 72 °Brix citrus molasses as a bioflavonoid complex. Citrus molasses contains approximately 2% hesperidin, which has been extracted with a solvent, concentrated, and recovered (Sokoloff, 1952). This process produces a dried or liquid concentrate of crude hesperidin and other flavonoids.

There has been some interest in recovering hesperidin from the caustic evaporator cleanup stream. The concentration of this caustic (2–5% NaOH) at the operating temperatures of the evaporator, necessary for efficient cleaning of the scale in the tubes, degrades the hesperidin. This degradation involves hydrolysis of the hesperidin sugar moiety to the aglycone, hesperetin, and is dependent on pH, concentration, and temperature. A portion of the hesperidin in the stream could probably be recovered if rapid pH adjustment to less than neutrality was performed as the solution exited the evaporator.

15.1.2 Properties

Powdered, crude hesperidin is nonhygroscopic, practically insoluble in water, and has a purity from 40 to 65%. Purified hesperidin has solubilities in water (20 ppm at 25°C, 75 ppm at 80°C), acetone (300 ppm at 50°C), ethyl acetate (150 ppm at 77°C), ethanol (0.14% at 78°C), methanol (0.28% at 25°C), isopropanol (600 ppm at 82°C), and dimethyl formamide (25% at 22°C). Pure hesperidin [molecular weight (MW) 610.6] has a melting point of 261°C and crystallizes as needles from dilute acetic acid. The sugar linkage contains a molecule of rutinose (rhamnose attached to glucose at the 6-position), which is attached to the hesperetin molecule (Fig. 15.1). Hydrolysis in alcoholic mineral acid converts hesperidin to the aglycone, hesperetin (Wingard, 1979). Alkaline solution favors the chalcone, which forms bright yellow crystals. Sour orange contains neohesperidin, with the sugar neohesperidose (rhamnose + glucose attached at the 2-position) (Fig. 15.1). Flavanone rutinoses are tasteless, while the neohesperidoses are bitter. Discussion and structures of these molecules have been reviewed (Horowitz and Gentili, 1977).

Hesperidin causes defect problems in concentrated juice manufacture, as the crystals precipitate in the evaporators, break free, and appear as small whitish-yellow flakes in the juice. Prevention is by good maintenance of evaporators and heat exchangers. This problem has been especially troublesome in pulp wash concentrate (Baker and Tatum, 1986).

15.1.3 Utilization

There has been considerable comment about the medical and pharmacological uses of hesperidin and other citrus flavonoids; however, dose-related responses have yet to be proven for these compounds in human nutrition. Much of the proposed utilization of hesperidin has been in difficult-to-prove applications, largely related to permeability and fragility of blood vessels and capillaries. There have been some patents for preparation of potentially useful chemical derivatives of hesperidin. These include conversions to sweet tasting chalcones such as hesperetin dihydrochalcone (Horowitz and Gentili, 1977) and to azo

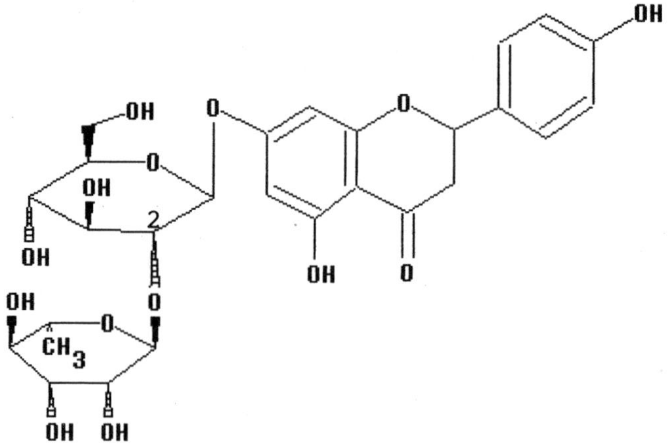

(*a*)

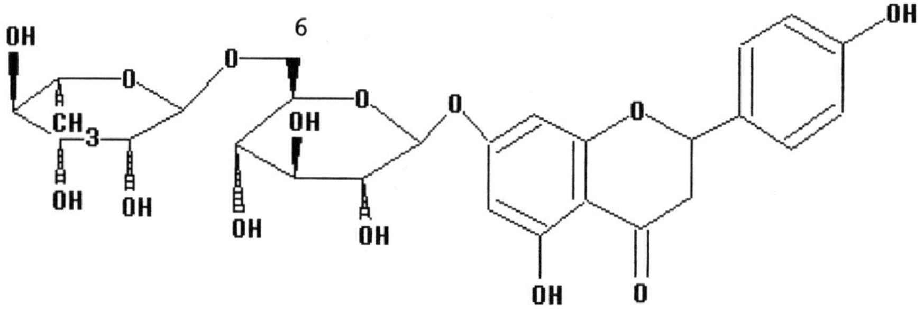

(*b*)

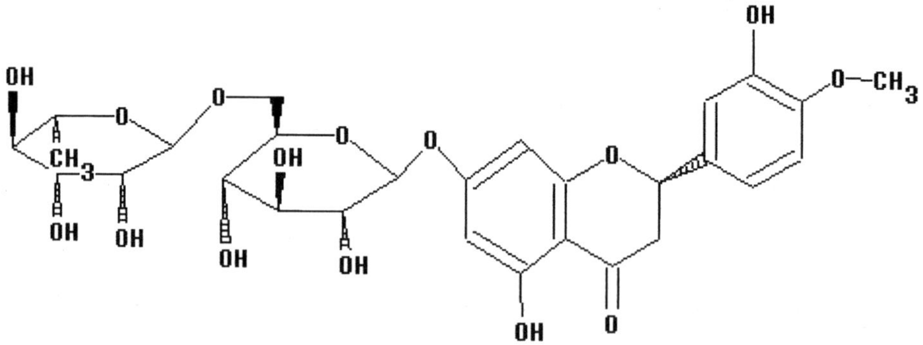

(*c*)

Figure 15.1 Chemical structures of (*a*) naringin (bitter), (*b*) narirutin (nonbitter), and (*c*) hesperidin (nonbitter).

dyes (Hendrickson and Kesterson, 1956a). Some further comments about medicinal uses of flavonoids are presented in Chapter 17.

15.2   NARINGIN

The bitter glucoside, naringin, found in grapefruit and pummelo distinguishes these fruits from other citrus. Naringin is chemically related and distributed in the fruit components in the same manner as hesperidin (Table 15.1) and likewise decreases with increasing maturity. This glucoside has been the subject of considerable research and, although not as pharmacologically interesting as other citrus flavonoids, it has by-product potential.

15.2.1   Recovery

Naringin can be extracted from the peel under alkaline conditions with lime and/or caustic similar to hesperidin (Baier, 1947). Since it is more water soluble than hesperidin, a lower pH of 9–10 is used. Precipitation and recovery of the crystals are similar to that described above, with solution concentration requirements needing to be higher than hesperidin to allow good crystallization.

A process, which avoids use of alkaline solutions, by extracting naringin from grapefruit residue with hot water was studied. After separation of the naringin extract, the residue was a suitable raw material for pectin extraction (Poore, 1934). Recent commercialization of adsorbent resins to debitter pulp wash and grapefruit juice (Chapter 6) has suggested the recovery of naringin and other flavonoids from caustic regeneration of the resin. While naringin was easily removed from process streams by adsorbent resins, recovery from the hot, 2% caustic solutions used to clean and regenerate the resins was in the range of 20% of the amount removed (author's unpublished data). The hot caustic degraded the naringin to the aglycone, naringinin, similar to caustic cleaning of hesperidin scale from evaporators. Use of alcohol or methanol to clean the resin allowed 80–90% naringin recovery; however, this adds expense to a commercial process.

15.2.2   Properties

Purified naringin ($C_{27}H_{32}O_{14} \cdot 2H_2O$, MW = 616.57) crystallized from alcohol or acetone and dried at 110°C has a melting point of 171°C. When crystallized from water, with an extra six molecules of bound water, the melting point is 83°C. Naringin solubility varies with water temperature at values of 0.17 g/L, 6°C; 0.5 g/L, 20°C; 1.96 g/L, 45°C; 7.16 g/L, 55°C; and 108 g/L, 75°C (Pulley, 1936). Crude, commercial naringin may be easily purified by dissolution (8–10% naringin) in hot (boiling) isopropanol and recrystallizing (Hendrickson and Kesterson, 1956b). The crystals form needles, which are easily broken. Naringin reacts with iron to form a black colored coordination complex in high concentration (red in low concentration). Solutions (0.5–1.0%) at pH 9.0 are quite viscous, 2000–5000 cP (Kesterson and Hendrickson, 1953). Naringin has the sugar, neohesperidose (glucose + rhamnose at 2-position), which confers bitterness; while the nonbitter isomer of naringin, narirutin, contains rutinose (glucose + rhamnose at 6-position) (Fig. 15.1). Refluxing naringin in dilute mineral acid results in hydrolysis to naringinin, glucose, and rhamnose. Naringinin melts with decomposition near 248°C.

Analytical measurement of naringin in grapefruit by colorimetric procedures (Davis test) results in values for total naringin, which includes the nonbitter narirutin (Fig. 15.1). Narirutin does not contribute to juice bitterness. Most quality control laboratories now use HPLC as a method of choice to monitor naringin content of grapefruit juices because the true value of naringin can be measured, separate from the narirutin. The amount of narirutin in grapefruit juices is approximately 15–20% of the amount of naringin (Rouseff et al., 1987).

15.2.3   Utilization

Naringin presence in citrus juices usually indicates the presence of grapefruit juice; however, naringin has been identified in sour orange (*Citrus aurantium*) and some tangelo juices (Rouseff et al., 1987). This offers the possibility that naringin measurement alone may not be sufficient for adulteration detection of orange with grapefruit juice.

15.2.3.1 Chemical   Naringin has also been used as a raw material for the recovery of the sugar rhamnose by hydrolysis in 4% sulfuric acid, giving a crystalline yield of about 20% of the starting mass of naringin (Pulley and von Loesecke, 1939). The yield by this method was about 60% of the theoretical recovery. A similar acid hydrolysis additionally involves selective cooling to yield a semisolid naringenin-*O*-glucoside and solution, from which rhamnose may be isolated (Kratky and Tandy, 1991). Rhamnose has also been prepared from naringin by an enzyme hydrolysis (Cheetham and Quail, 1991). The aglycone, naringenin, can be hydrolyzed under basic conditions to yield *p*-coumaric acid and phlorglucinol, starting materials for other organic syntheses (Kesterson and Hendrickson, 1953). Naringin has also been studied for its antioxidant properties, and tested better than butylated hydroxy toluene (BHT) to prolong vegetable oil stability (Das and Pereira, 1990).

15.2.3.2 Bitter Flavoring   Due to its bitterness (20 ppm threshhold in water) and solubility, naringin has been used in formulations to provide bitterness to foods and beverages. It has been suggested as a replacement for caffeine and quinine in bitter marmalade, tonic water, and grapefruit-flavored products (Anon., 1982). Citrus flavored desserts and other foods may have their mouthfeel adjusted by adding 0.1–50 ppm of a bittering agent containing naringin (Soukup and Parliament, 1984).

15.2.3.3 Sweetener   The sweeteners, naringin dihydrochalcone and neo-hesperidin dihydrochalcone are, respectively, 100 and 1500 times sweeter than sucrose. Both may be chemically synthesized from naringin, although the interest centers on the latter because of its intense sweetness (Robertson et al., 1974). Successfully formulating flavors with this sweetener requires careful use of bulking agents to avoid lingering aftertaste, and each product must be examined individually for its suitability (Pratter, 1980). For example, these sweeteners can delay deterioration of essential oil flavors in chewing gum, in addition to providing sweetness in the absence of sugar (Westall et al., 1974). A thorough review of neohesperidin dihydrochalcone properties and worldwide food and beverage applications has been published (Bär et al., 1990).

15.3 POLYMETHOXYLATED FLAVONES

The most common polymethoxylated flavones—nobiletin, sinensetin, and tangeretin—of citrus have been the subject of interest for their reported pharmacological properties. These properties have been suggested to be related to chemicals that affect the adhesive properties of blood cells (Robbins, 1975). The concentration of these flavonoids in citrus is much less than that of hesperidin and naringin, although the amounts in peel oil and wax are significant. Polymethoxylated flavones were made in commercial quantities by extraction from citrus molasses and sold as a therapeutically active product, vitamin P (Sokoloff, 1952).

The chemistry, properties, and occurrence in the various citrus fruits of polymethoxylated flavones has been reported (Horowitz and Gentili, 1977). Information of their distribution in the component parts and by-products of fruit is presented in Table 15.2. Highest concentrations occur in the wax isolated from cold-pressed peel oil. Most of the amount in the peel is present in the flavedo or peel oil (Swift, 1960). For example, Dancy tangerine peel, which

TABLE 15.2   Amounts of Polymethoxylated Flavones in Components and By-products of Oranges

Component	Nobiletin (ppm)	Sinensetin (ppm)	Tangeretin (ppm)
Peel[a,b]	300–500	200–300	5–100
Juice sacs[a]	<1	<1	<0.2
Seed[a]	<1	<1	<0.2
50 °Brix molasses[b]	500–700	300–500	50–80
C.P. oil[b]	2,700	400	800
Wax[b]	17,000	2,000	13,000
Juice[a]	1	<1	<0.2

[a]Rouseff and Ting (1979b).
[b]Manthey and Grohman (1996).

has no albedo, contains almost 200 ppm tangeretin (Rouseff and Ting, 1979a). If unwaxed, cold-pressed, or folded oils are held for a long time in cold storage, it is common for these compounds to precipitate. The internal, edible parts of fruit, the juice, and pulp contain only trace amounts of polymethoxylated flavones.

15.4 LIMONOIDS

The tetracyclic triterpenoid compounds, known as limonoids, present in citrus seeds and tissue of fruit have little commercial value. However, similar to flavonoids, there has been recent interest in specific pharmacological properties of these compounds from the medical community.

15.4.1 Limonin

The intensely bitter compound, limonin, is notably responsible for the bitterness of navel oranges and with naringin, grapefruit juice (Chapter 6). There have been many studies of the properties of this compound and ways to remove it from juice, but few reports of methods to isolate it, other than for analytical purposes (Chandler et al., 1976; Chandler and Robertson, 1983).

15.4.1.1 Recovery   Limonin, like the flavonoids, varies in concentration within the fruit tissues, with the highest percentage in the seeds (0.5–1.0% dry wt.), tissue (0.05–0.2%) with lower amounts in the juice (1–50 ppm) (Chandler and Robertson, 1983).

A procedure to prepare gram quantities of limonin from citrus seeds has been used in the author's laboratory as follows: Wet seeds from a grapefruit sectionizing operation were dried to 8% moisture and pulverized to a meal in a Fitzpatrick hammer mill. Extract the lipid in the meal (1 kg) by grinding in a blender with hexane (2 L), filter to recover meal, discard extract, and repeat twice with 700 mL hexane. (CAUTION! Hexane is flammable.) Air dry to remove hexane from the meal. Extract the limonin by grinding the defatted meal with dichloromethane (700 mL), repeat twice with 500 mL. Filter to recover the filtrate. Evaporate the solvent to dryness to recover the crystals. Wash the crystals with hexane to remove residual fat. Yield of crude limonin from 1 kg dried seeds is approximately 10 g. The crude limonin may be recrystallized from dichloromethane to increase purity.

15.4.1.2 Properties   High-purity limonin ($C_{26}H_{30}O_8$, MW = 470.52) forms white crystalline flakes with a very clean, delayed bitterness when tasted (Merck, 1996). Water solubility is approximately 5 mg/L, while in bitter citrus juices may be as high as 40 mg/L (Chandler and Robertson, 1983). The sensory threshold is in the range of 3–6 mg/L. The chemistry and properties of limonin in its relation to juice bitterness has been reviewed (Maier et al., 1980) and related to the structural properties of the bitter and nonbitter forms (Fig. 6.2).

15.4.2 Limonoid Glucosides

The current interest in natural chemopreventive substances has stimulated research of limonoid glucosides and their properties. The concentration of these compounds increases in fruit tissue until maturity, reaching about 40–50 mg/fruit for Valencia oranges (Fong et al., 1992). For reference, this may be compared to about 1 g hesperidin/fruit in oranges.

15.4.2.1 Recovery There is no commercial recovery procedure for limonoid glucosides, as there is no economic use for them. Extractions from fruit tissues for analytical chemical quantities and small amounts for experimental studies have been performed with solvents such as methanol. In general, the tissue is blended with methanol, centrifuged, and the residue re-extracted with 70% methanol, filtered, passed through a nonpolar adsorbent, which was rinsed with water and then methanol to elute the limonoid glucoside (Ozaki et al., 1995; Ifuku et al., 1998). These authors also isolated this material from Satsuma molasses, while others have recovered it from orange molasses. Molasses from commercial orange juice processing residue contained approximately 2 g/L limonoid glucosides at 40 °Brix (Hasegawa et al., 1996). This amount is considerably more than the polymethoxylated flavones in molasses (Table 15.2), but much less than the amount (2%) of flavonoids in 72 °Brix molasses (Sokoloff, 1952). Grapefruit seeds have been shown to contain about 1% of these compounds (Hasegawa et al., 1989). Unless limonoid glucosides become extremely valuable, recovery processes will probably not be economical.

15.4.2.2 Properties The limonoid glucosides are somewhat soluble in water and can be extracted from the fruit tissue. Chemical isolation from citrus seeds has resulted in identification and nuclear magnetic resonance (NMR) characterization of several of these compounds (Hasegawa et al., 1989), of which limonin-17-β-D-glucoside is the most common (Fig. 6.1). There is only speculative use for these compounds, based on their reputed biological activity against tumors and as insect antifeedants (Hasegawa et al., 1996).

REFERENCES

Anon. (1982). Naringin, a bitter flavoring, available on high-purity grades, *Food Product Dev.* **16**(1):29.

Baier, W. E. (1947). Methods for recovery of naringin, U.S. Pat. 2,421,063.

Baker, R. A. and Tatum, J. H. (1986). Hesperidin precipitation from orange pulp wash, *Proc. FL State Hort. Soc.* **99**:90–92.

Bär, A., Borrego, F., Benavente, O., Castillo, J., and del Rio, J. A. (1990). Neohesperidin dihydrochalcone: Properties and applications, *Lebensm.-Wiss. u.-Technol.* **23**(5):371–376.

Chandler, B. V. and Robertson, G. L. (1983). The solubility of limonin, the bitter principle of orange juice, *J. Sci. Food Agric.* **84**:1272–1284.

Chandler, B. V., Nicol, K. J., and von Biedermann, C. (1976). Factors controlling the accumulation of limonin and soluble constituents within orange fruits, *J. Sci. Food Agric.* **27**:866–876.

Cheetham, P. S. J. and Quail, M. A. (1991). Process for preparing L-rhamnose, U.S. Pat. 5,077,206.

Das, N. P. and Pereira, T. A. (1990). Effects of flavonoids on thermal autoxidation of palm oil: Structure-activity relationships, *J. Am. Oil Chem. Soc.* **67**(4):255–258.

Davis, W. B. (1947). Determination of flavanones in citrus fruit, *Anal. Chem.* **19**:476–478.

Fong, C. H., Hasegawa, S., Coggins, C. W., Jr., Atkin, D. R., and Miyake, M. (1992). Contents of limonoids and limonin 17-β-D-glucopyranoside in fruit tissue of Valencia orange during fruit growth and maturation, *J. Agric. Food Chem.* **40**(7):1178–1181.

Hasegawa, S., Bennett, R. D., Herman, Z., Fong, C. H., and Ou, P. (1989). Limonoid glucosides in citrus, *Phytochemistry* **28**(6):1717–1720.

Hasegawa, S., Fong, C. H., Miyake, M., and Keithly, J. H. (1996). Limonoid glucosides in orange molasses, *J. Food Sci.* **61**(3):560–561.

Hendrickson, R. and Kesterson, J. W. (1954). Recovery of citrus glucosides, *Proc. FL State Hort. Soc.* **67**:199–203.

Hendrickson, R. and Kesterson, J. W. (1956a). Glucoside azo dyestuffs, U.S. Pat. 2,748,107.

Hendrickson, R. and Kesterson, J. W. (1956b). Purification of naringin, *Proc. FL State Hort. Soc.* **69**:149–152.

Hendrickson, R. and Kesterson, J. W. (1964). Hesperidin in Florida oranges, FL Agr. Exp. Sta. Tech. Bull. No. 683, University of Florida, Gainesville, FL.

Higby, R. H. (1947). Methods for recovery of flavanone glucosides, U.S. Pat. 2,400,693.

Horowitz, R. M. and Gentili, B. (1977). "Flavonoid constituents of citrus." In *Citrus Science and Technology*, Vol. 1, S. Nagy, P. E. Shaw, and M. K. Veldhuis, Eds., AVI, Westport, CT, Chap. 10, pp. 110–207.

Ifuku, Y., Maeda, H., and Miyake, M. (1998). Method for manufacturing limonoid glucosides, U.S. Pat. 5,734,046.

Kesterson, J. W. and Hendrickson, R. (1953). Naringin, a bitter principle of grapefruit, FL Agr. Exp. Sta. Tech. Bull. No. 511, University of Florida, Gainesville, FL.

Kratky, Z. and Tandy, J. S. (1991). Selective cleavage of naringin, U.S. Pat. 5,008,381.

Maier, V. P., Hasegawa, S., Bennett, R. D., and Echols, L. C. (1980). "Limonin and limonoids. Chemistry, biochemistry and juice bitterness." In *Citrus Nutrition and Quality*, S. Nagy and J. A. Attaway, Eds., ACS Symp. Ser. 143, American Chemical Society, Washington, D.C., pp. 63–82.

Manthey, J. A. and Grohmann, K. (1996). Concentrations of hesperidin and other orange peel flavonoids in citrus processing byproducts, *J. Agric. Food Chem.* **44**(3):811–814.

Merck (1996). *The Merck Index on CD-Rom*, 12th ed., Version 12:1, Chapman & Hall, New York.

Ozaki, Y., Ayano, S., Inaba, N., Miyake, M., Berhow, M. A., and Hasegawa, S. (1995). Limonoid glucosides in fruit, juice and processing by-products of Satsuma Mandarin (*Citrus unshiu* Marcov.), *J. Food Sci.* **60**(1):186–189, 194.

Poore, H. D. (1934). Recovery of naringin and pectin from grapefruit residue, *Ind. Eng. Chem.* **26**(6):637–639.

Pratter, P. J. (1980). Neohesperidin dihydrochalcone: An updated review on a naturally derived sweetener and flavor potentiator, *Perf. Flavorist* **5**:12–14, 16–18.

Pritchett, D. E. and Merchant, H. E. (1946). The purification of hesperidin with formamide, *J. Am. Chem. Soc.* **68**:2108–2109.

Pulley, G. N., and von Loesecke, H. W. (1939). Preparation of rhamnose from naringin, *J. Am. Chem. Soc.* **61**(1):175–176.

Robbins, R. C. (1975). Compositions and methods for disaggregating blood cells, U.S. Pat. 3,903,266.

Robertson, G. H., Clark, J. P., and Lundin, R. (1974). Dihydrochalcone sweeteners: Preparation of neohesperidin dihydrochalcone, *Ind. Eng. Chem., Prod. Res. Dev.* **13**(2):125–129.

Rouseff, R. L. and Ting, S. V. (1979a). Tangeretin content of Florida citrus peel as determined by HPLC, *Proc. FL State Hort. Soc.* **92**:145–148.

Rouseff, R. L. and Ting, S. V. (1979b). "Analysis of polymethoxylated flavones in orange juice and fruit parts." In *Liquid Chromatographic Analysis of Food and Beverages*, Vol. 2, Academic, New York, pp. 537–558.

Rouseff, R. L., Martin, S. F., and Youtsey, C. O. (1987). Quantitative survey of narirutin, naringin, hesperidin and neohesperidin in citrus, *J. Agric. Food Chem.* **35**(6):1027–1030.

Sokoloff, B. T. (1952). Method of extracting vitamin P from citrus molasses, U.S. Pat. 2,734,896.

Soukup, R. J. and Parliament, T. H. (1984). Flavor and mouthfeel character in foodstuffs by the addition of bitter principles, U.S. Pat. 4,479,972.

Swift, L. J. (1960). Nobiletin from the peel of the Valencia orange (Citrus sinensis L.), *J. Org. Chem.* **25**:2067–2068.

Szent-Gyorgyi, A. (1938). Methoden zur herstellung von Citrin [Preparation of citrin], *Zeitschrift f. Physiol. Chem.* **255**:126–131; *Chem. Abst.* **32**:9393 (1938).

Westall, E. B., Scanlan, J. J., and Sahaydak, M. (1974). Flavor preservation in sugarless chewing gum compositions and candy products, U.S. Pat. 3,857,962.

Wingard, R. E. (1979). Conversion of hesperidin into hesperetin, U.S. Pat. 4,150,038.

CHAPTER 16

SEED PRODUCTS

Current industrial practice is to include the seeds in the juice processing residue for the production of dried cattle feed. However, since the early 1900s, citrus seeds have been recognized as a source of edible oil for human food use, with its components of palmitic, stearic, oleic, linoleic, and linolenic acids (Diedrichs, 1914). The oil from lemons and sour oranges was characterized chemically and physically, and the intense bitter taste of limonin was recognized (Peters and Frerichs, 1902). At present, no citrus seed oil is produced commercially, the last operation ceasing production in Florida after the 1970 season (Braddock and Kesterson, 1973). Because the common desirable juice oranges (Hamlin and Valencia) have few seeds and consumers prefer seedless grapefruits (few seedy grapefruits remain), viable commercial seed recovery from processing residue is questionable. However, interest in citrus seeds continues, and a review (56 references) of compiled technical and chemical data is not widely accessible (Braddock and Kesterson, 1973).

16.1 RECOVERY

16.1.1 Seeds

The quantity of seeds in some citrus fruit and corresponding amounts in the dried peel residue are presented in Table 16.1. Separation of seeds from the rag, pulp, and peel residue may be accomplished by dropping the material into a tilted rotating reel fixed with a coarse screen, allowing the smaller pieces of residue and seeds to fall through the openings, where seeds are recovered by a paddle finisher. Alternatively, the dried residue may be passed through a series

TABLE 16.1   Citrus Seed Percentage in Oranges, Grapefruits, and Dried Pulp

Variety	Fruit % (w/w)[a]	Dry Pulp % (w/w)[a]
Hamlin	0.5	1.6
Pineapple	4	12
Valencia	1	3
Dancy	1	3
Murcott	3	12
Temple	2	8
Duncan	5	16
Marsh	0.5	2
Ruby	0.5	2

[a]Whole fruit percentage based on wet weight, dried pulp at 10% moisture.

of cyclone separators, screens, and winnowing devices, recovering seeds based on density (USDA, 1956). Seed separation may also be accomplished by hand from citrus fruit sectionizing operations. A novel seed collection procedure involves screw conveyor transfer of the peel residue to a scrambling device, which throws the seed-containing material onto a slanted moving conveyor belt. The residue sticks to the belt and the seeds bounce, striking a baffle, and are diverted to a storage bin (Kirk, 1967). Recovered wet seeds were limed and stored until dehydration in a citrus peel dryer.

16.1.2 Seed Oil

The dried seeds have to be accumulated from a number of citrus processing plants in order to have a large enough quantity to operate a seed oil recovery plant. This is a logistical problem of considerable magnitude. Traditional oil recovery was by expelling (pressing) oil from the intact dried seeds (8–10% moisture), without removing the hulls (Nolte and von Loesecke, 1940; Van Atta and Dietrich, 1944). The recovery of oil by expelling is about 25–30% of the dry seed mass (500–600 lb oil/ton seeds), leaving 2–5% oil in the press cake (seed meal).

Following recovery, the crude oil must be refined in the manner of traditional oil seeds for edible oil use, as it contains free fatty acids, phosphatides, pigments, and bitter limonoids. Free fatty acid content of citrus seed oils has been determined to be in the range of 0.3–1.4% (Braddock and Kesterson, 1973). Refining is commonly performed by treatment of the crude oil with a slight excess of alkali over the amount required to react with the free fatty acids in the oil.

16.2 COMPOSITION

16.2.1 Seed Oil

The most notable feature of citrus seed oil is the large proportion of polyunsaturated fatty acids in the triglycerides. General ranges for the fatty acid composition of citrus seeds are listed in Table 16.2. The high contents of oleic, linoleic, and linolenic acids allow the oil to be liquid at room temperature. Other edible oil constants have been reviewed for seed oil and include the values for specific gravity (0.9170), refractive index (1.4698), acid value (0.95), iodine value (100.9), saponification value (193), acetyl value (2.4), Reichert–Meissl number (0.47), and Polenske number (0.2) (Nolte and von Loesecke, 1940; Braddock and Kesterson, 1973). Citrus seed oil also contains from 30 to 60 mg/100 g of vitamin E.

16.2.2 Seed Meal

After expelling the oil, the protein content of the residual components was meal (22%) (Nolte and von Loesecke, 1940), dried whole seeds (16%), kernels (19.5%), and hulls (6.1%) (Ammerman et al., 1963). The amino acid composition of citrus seed meal listed in Table 16.3 indicates this product is of good nutritional quality for livestock feeding, specifically for the values of cystine, glycine, methionine, and tryptophan.

A toxic factor was reported in citrus seed meal fed to chicks as 20% of the diet (Driggers et al., 1951). This factor was thought to be limonin and resulted in unsatisfactory growth and high mortality.

16.3 UTILIZATION

Refined citrus seed oil is pale yellow and has a bland flavor, similar to other edible oils. It readily becomes rancid because of the concentration of linoleic

TABLE 16.2 Fatty Acid Composition of Citrus Seeds

Variety	Percent of Total					
	Palmitic	Palmitoleic	Stearic	Oleic	Linoleic	Linolenic
Orange[a,b]	26–31	0.1	3–5	24–28	35–37	2–4
Grapefruit[c]	26–36	0.1–0.3	1–4	18–25	32–40	3–6
Mandarin[d]	22–30	0.1–1	2–5	20–25	37–45	3–5
Lemon[b,c]	20–24	0.1–0.3	2–4	26–31	31–38	8–12
Lime[b,c]	24–29	0.1–0.5	3–5	20–22	37–40	6–11

[a]Hendrickson and Kesterson (1963).
[b]Nordby and Nagy (1969).
[c]Braddock and Kesterson (1973)
[d]Hendrickson and Kesterson (1964).

TABLE 16.3   Amino Acid Composition of Citrus Seed Meal[a]

Amino Acid	mg Amino Acid/g Nitrogen[a]	
	Orange	Grapefruit
Aspartic	548	560
Threonine	186	181
Serine	239	290
Glutamic	1594	1623
Proline	256	253
Glycine	322	272
Alanine	231	230
Cystine	110	174
Valine	307	333
Methionine	112	165
Isoleucine	219	224
Leucine	394	446
Tyrosine	168	166
Phenylalanine	306	296
Lysine	178	175
Histidine	128	108
Arginine	695	596
Tryptophan	1254	79

[a]Adapted from Braddock and Kesterson (1972).

and linolenic acids in the triglycerides. It has been hydrogenated and used in margarine (Anon, 1964). A mixture of the oil and defatted meal has also been proposed as a natural beverage clouding agent (Kesterson et al., 1969). Grapefruit seed oil has been suggested as a lubricant and preservative to treat textile fibers (Kaplan, 1941) and in paint and soap formulations (Stambovsky, 1942).

Concern that removal of the seeds from the dried pulp cattle feed would significantly reduce the protein and fat content has been studied (Braddock and Kesterson, 1973). Since few seedy varieties are now processed, the reduction would not be statistically significant, should seed oil processing be revived.

REFERENCES

Anon. (1964). Margarine from citrus seed oil, *Chemurgic Digest* **22**(4):1.

Ammerman, C. B., van Wallegham, P. A., Easley, J. F., Arrington, L. R., and Shirley, R. L. (1963). Dried citrus seeds—nutrient composition and nutritive value of protein, *Proc. FL State Hort. Soc.* **76**:245–249.

Braddock, R. J. and Kesterson, J. W. (1972). Amino acids of citrus seed meal, *J Am. Oil Chemists' Soc.* **49**:671–672.

Braddock, R. J. and Kesterson, J. W. (1973). Citrus seed oils, FL Agr. Exp. Sta. Tech. Bull. No. 756, University of Florida, Gainesville, FL.

Diedrichs, A. (1914). Über bisher wenig untersuchte Samen und deren Öle. 2. Apfelsinen und Citronensamen, *Z. Untersuch. Der Nahrung. Genussmittel.* **27**:132–135.

Driggers, J. C., Gavis, G. K., and Mehrhof, N. R. (1951). Toxic factor in citrus seed meal, FL Agr. Exp. Sta. Tech. Bull. No. 476, University of Florida, Gainesville, FL.

Hendrickson, R. and Kesterson, J. W. (1963). Seed oils from Citrus sinensis, *J. Am. Oil Chemists' Soc.* **40**:746–747.

Hendrickson, R. and Kesterson, J. W. (1964). Seed oils from Florida mandarins and related varieties, *Proc. FL State Hort. Soc.* **77**:347–351.

Kaplan, P. (1941). Treating textile fibers such as silk, wool and rayons, U.S. Pat. 2,229,975.

Kesterson, J. W., Hendrickson, R., and Adkins, C. D. (1969). Citrus seed clouding agent for beverage bases, U.S. Pat. 3,660,105.

Kirk, W. A. (1967). U.S. Pat. 3,330,410.

Nolte, A. J . and von Loesecke, H. (1940). Grapefruit seed oil: Manufacture and physical properties, *Ind. Eng. Chem.* **32**:1244–1246.

Nordby, H. E. and Nagy, S. (1969). Fatty acid profiles of citrus juice and seed lipids, *Phytochemistry* **8**:2027–2038.

Peters, W. and Frerichs, G. (1902). Über das fette Oel der Zitronenkerne und das Limonin, *Archiv. Pharmazie.* **240**:659–662.

Stambovsky, L. (1942). Citrus seed oil, *Drug Cosmetic Ind.* **51**:156–157.

USDA (1956). Chemistry and technology of citrus, citrus products and by-products, U.S.D.A./A.R.S. Agriculture Handbook No. 98, **17**:60–61.

Van Atta, G. R. and Dietrich, W. C. (1944). Valencia orange seed oil, *Oil Soap* **21**(1): 19–22.

CHAPTER 17

NUTRACEUTICALS, MYSTICAL SUBSTANCES, AND BY-PRODUCTS

Plant-derived substances have long held mystic in human nutrition because many common pharmaceuticals are based on chemicals found in nature. When useful chemicals from natural sources are scrutinized scientifically and found to be healthful, some may profit from marketing claims about such chemicals or food products containing them. This may be true even if the healthful compound is present in trace quantities, below any effective concentration for metabolic reactions when the food is consumed. Realistic examples include vitamins, certain antioxidants, micronutrients, and a few drugs. Natural chemicals that are bad or toxic are also common in the human diet. Examples include the toxic alkaloids, photosensitizing coumarins, antidigestive factors, and microbial toxins. Presence of trace amounts of these chemicals in a popular food may be ignored or not mentioned because of the susceptibility of the public to adverse media information.

Because of public interest in potential health benefits of such substances, partially stimulated by advertising, this chapter briefly addresses some nutritional and medical aspects of certain natural substances in citrus juices and by-products.

17.1 VITAMINS AND NUTRIENTS

Nutritionally, citrus and other fruits contain mostly water and carbohydrates, with lesser quantities of other chemicals. There are valid nutritional benefits for a few of the vitamins and minerals; however, there are unproven claims for some of the chemicals present naturally in citrus juices and products.

17.1.1   Vitamin C and Folate

The traditional consumption of citrus juices as a breakfast beverage has historical ties to obtaining one's daily requirement of vitamin C. Citrus and many other fruits are good sources of this vitamin (Table 17.1) and there are several hundred references on this subject. From a by-products viewpoint, however, it would not be economical to consider extraction and recovery of a substance whose concentration in the fruit was only 50 mg/100 g. The amount of vitamin C in 8 fluid oz. of orange juice is usually between 100 and 200% of the daily value based on a 2000-calorie diet. The juice extraction residues of oranges and grapefruit also have been shown to contain over 70% of the vitamin C of the whole fruit (Atkins et al., 1945).

Folic acid has received considerable media attention because of its dietary requirement early in pregnancy to prevent neural tube defects in infants (Roe, 1990). Current U.S. laws require listing the nutritional information on food labels. The daily nutritional value for folic acid is 400 μg, of which orange juice (8 oz.) contains approximately 10–20%, enough to be mentioned on product labels.

17.1.2   Other Nutrients

Orange juice (8 oz.) naturally contains some label amounts of a few other nutrients, specifically, portions of the daily values of thiamin (10%), and daily need of potassium (10–15%). There are also small amounts of other substances such as vitamin A, phosphorus, and magnesium. For marketing reasons, some manufacturers add certain nutrients, vitamins, and antioxidants to citrus juice products. The consumer may purchase for example, orange juice with added vitamins A, C, and E, niacin, and B_6, as well as additional calcium. The stan-

TABLE 17.1   Average Vitamin C Content of Some
Fresh Fruits and Vegetables

Product	mg/100 g
Guava	300
Green pepper	120
Cabbage	60
Orange	50
Lemon	50
Grapefruit	40
Tangerine	30
Tomato	25
Pineapple	25
Apple	10
Peach	4

dard reference for nutritional composition of juices is the online USDA nutritional database (USDA, 1998).

17.2 FLAVONOIDS

Flavonoids have dubious credentials of metabolic benefits in human nutrition as most experimental studies have been performed in vitro, and human doses are unknown. Traditional reference to citrus flavonoids as vitamin P still persists, although there is no evidence for the efficacy of these substances as a vitamin. Methods for screening mutagenicity of chemicals using bacteria (the Ames test) have shown that certain flavonoids, notably quercetin, exhibit positive test results (Horowitz, 1981). On the other hand, it has been published that quercetin inhibits growth of malignant intestinal cells (Attaway, 1994). Certainly, many foods contain this equivocal mutagen (red wine, tea, and grape juice), yet no dietary studies have linked these foods to cancer, and true scientific proof of their antitumor activity is lacking.

The Florida Department of Citrus has funded many research studies to relate health benefits of flavonoids and other components to dietary consumption of citrus juices and products. A number of these studies have been reviewed with interest in the effects of flavonoids, particularly polymethoxylated flavones, to prevent cancer, inhibit tumor development, or other pharmacolocical action (Attaway, 1994). One study indicated that human blood levels of the immunosuppressant drug, cyclosporin, increased in patients taking the drug orally mixed with grapefruit juice (Yee et al., 1995). It was suggested that flavonoids in the juice inhibited an intestinal enzyme, which degrades the drug, allowing more drug to enter the blood, and hence, smaller doses were required. While these reports are interesting, there is a lack of sound scientific studies to allow specific health claims under regulatory law. Practically, even if a dose was only a few milligrams, one would not recommend drinking juice to obtain the most popular of the studied compounds, polymethoxylated flavones, as the juice concentration is ≤ 1 ppm (Table 15.2).

17.3 LIMONOID GLUCOSIDES

Recent identification of limonoid glucosides in citrus molasses has offered a possible source of these chemicals for biological studies. Similar to the flavonoids, these compounds have reported activity against tumors (Hasegawa et al., 1996). Since the juice and fruit have very low amounts of these chemicals (50 mg/whole Valencia orange) (Chapter 15), and the seeds and peel tissue are very bitter, it is speculative to consider citrus to be a useful dietary source.

17.4   DIETARY FIBER

There is a large body of good evidence supporting the value of dietary fiber in human nutrition. As far as citrus is concerned, most of the fiber is in the peel, membranes, and rag and very little in the juice. Citrus juice contains about 10% pulp (juice sacs), which contains about 10 g fiber/100 g (wet wt.), allowing juice to contain about 1 g fiber/100 g juice. The peel and seeds contain over 10 times that amount. Also, citrus fiber is primarily pectin, which in dietary fiber jargon is considered soluble fiber, since it is solubilized during digestion. Clearly, to benefit from soluble citrus fiber one must eat the whole fruit, including the peel and seeds, or consume citrus pectin.

17.5   TRACE NATURAL CHEMICALS AS BY-PRODUCTS

For some of the above chemicals, the concentrations are too low to allow one to obtain effective amounts by drinking juice or eating the fruit, peel, seeds and all. The potential of extracting and recovering some of these natural compounds in citrus as useful drugs is an idea, which occasionally surfaces. This is also unlikely, for several reasons. For example, suppose a truly valuable chemical substance was present in peel residue at 0.1% wt/wt. The amounts of extracting solvent, distillation equipment, or other separation technology would have to be capable of dealing with the other 99.9% of the mass to recover the compound. Additionally, recovery efficiency would be less than 100%. Then, there is the issue of what to do with the recovery residue, which may now be in a form more difficult to handle than the original raw material. Finally, suppose the chemical had proven miraculous curative properties, as, for example, the cancer drug, Taxol, from the bark of yew trees. Then, organic chemists would simply synthesize it, and there would be no longer any need for citrus raw materials.

17.6   MYSTICAL SUBSTANCES FROM CITRUS

17.6.1   Useful Products

A quick search of the Internet will verify that there are hundreds of useful products and substances with mystical properties with their origins in citrus by-products. The author, not wishing to squelch creativity with references, will allow readers to explore at their leisure. There are also many patents for helpful products to be made from citrus substances. Here are some of the best. Should one overindulge, probably from reading the profound content of this text, there is a hangover-reducing product that contains up to 85% hesperidin (Moldowan and Moldowan, 1985). A nutritional supplement that contains hesperidin and vitamin C can enhance one's immune system to prevent infection (Paul, 1997).

Citrus bioflavonoid complex is part of a composition that reduces crises from sickle cell anemia (Lockett, 1997). Naringin can selectively inhibit growth of cancer cells that have resistance to chemotherapy or radiation (Beljanski, 1992). One might wonder whether doses of bitter naringin could be considered chemotherapy.

Hesperidin seems to have many potential uses. One patent suggests that mouthwash preparations containing the active ingredient hesperidin effectively treats gingivitis (Raaf, 1985). A derivative of hesperidin, α-glycosl hesperidin, is useful as a component of vitamin P agents, pet foods, tobacco, pharmaceuticals, cosmetics, and plastics (Hijiya and Miyake, 1997). Bioflavonoids also have antimicrobial properties (Frazier, 1980) and help milk preserve freshness if added before pasteurization (Morgan et al., 1971). A mixture containing hesperidin also has been shown to reduce bleeding after a hemorrhoidectomy (Foo et al., 1995). There are many more applications not listed here.

17.6.2   Mystical Seed Products

Besides their usefulness for edible oil products, citrus seeds are probably ranked second, only to the flavonoids, for the most product applications. Most of these applications utilize a "grapefruit seed extract," a mostly unidentified substance. The most famous of these products, DF-100, has reportedly wide-ranging antimicrobial properties effective against a broad range of human pathogens, fungi, and food spoilage organisms (Harich, 1997). It is also an effective antioxidant and can help preserve fish filets in cold storage (Cho et al., 1990). DF-100 and another citrus seed extract, Citrex, also claim to have antifungal and preservative properties for postharvest treatments of citrus and other fruit (Cho et al., 1993; Schirra and Mulas, 1995).

There are even books describing the uses of grapefruit seed extracts for the treatment of numerous maladies that cause misery for humankind (and citrus by-product experts). The list is long, but grapefruit seed extracts have curative powers for many infections of the scalp, skin, and even nail fungus. This wonderful agent claims to cure vaginal yeast infections, gastric upset, ulcers, and kill many human parasites. Commercial uses mainly relate to its disinfectant properties, in swimming pools to replace use of high concentrations of chlorine, in livestock drinking water, and to kill mites, aphids, and mildew of plants (Sharamon and Baginski, 1996; Sachs, 1997).

These products are not without problems since safety issues, adulteration with chemical fungicides, and false claims are recognized. One study of the composition of DF-100 reported finding the presence of the antimicrobial chemicals, methyl-p-OH-benzoate and 2,4,4'-trichloro-2'-OH-diphenyl ether (Nishina et al., 1991). Certainly, neither of these chemicals are natural components of grapefruit seeds. Many common food spoilage organisms have been found to develop resistance to grapefruit seed extracts, even at high use levels of 1–5% (Sundheim and Langsrud, 1995). Commercial grapefruit seed extract caused moderate erythema to the skin and severe eye injury applied to rabbits

(Ko et al., 1995). A food additive petition was filed with the FDA proposing that grapefruit seed extract be permitted as an antimicrobial agent in processing fresh or frozen poultry, fish, and shellfish (Federal Register, 1995). Later claims of false data in this petition were published, claiming the product contained the synthetic chemicals triclosan, methylparaben, and benzethonium chloride (Food Chemical News, 1995).

The author provides this last comment. Florida grows and processes over 50% of the world's grapefruit, and grapefruit must be processed to obtain seeds. There is currently no large-scale processing recovery of grapefruit seeds (or other citrus) at citrus plants, particularly since the last seed oil manufacturer ceased business in 1970 (Braddock and Kesterson, 1973). The author would like to know the source of the seeds used in the manufacture of grapefruit seed extracts.

17.7 BY-PRODUCTS

From the discussions of products described in the various chapters, the viable citrus by-products produced commercially and their yields are listed in Table 17.2. The nature of the raw material limits the useful products to these and a few others, which imaginative minds may still create. At this point it is also

TABLE 17.2 Potential Yield of By-products from Oranges and Grapefruits

Product	Yield (kg/box)[a]	
	Oranges	Grapefruit
Dry pulp w/molasses (10% H_2O)	4.0	4.1
Dry pulp w/o molasses	2.9	2.8
Molasses (72 °Brix)	1.4	1.5
Ethanol from molasses[b]	0.2	—
Cold-pressed oil and limonene	0.3	0.1
Pulp wash solids	0.2–0.3	0.1–0.2
Frozen pulp	2	3
Oil-phase essence	0.005	0.003
Aqueous essence at 15% alcohol	0.02	0.02
Pectin (150 grade)	1.3	1.0
Seed oil	0.05	0.09
Hesperidin	0.2	—
Naringin	—	0.2

[a]1 box oranges = 40.9 kg, juice yield = 55%; 1 box grapefruit = 38.6 kg, juice yield = 48%.
[b]The value is in L/box of 100% ethanol.

edifying to review the studies cited in Chapter 1 to see how far we have come since the first citrus by-products (Timmons, 1950; Will, 1916). There are other products not listed in Table 17.2, such as cloud, peel fiber, and some specialty essential oils; however, these are difficult to relate to the mass of fruit processed. The reader is referred to the specific chapters for technical manufacturing information of these products.

REFERENCES

Atkins, C. D., Wiederhold, E., and Moore, E. L. (1945). Vitamin C content of processing residue from Florida citrus fruits, *Fruit Prods. J. and Amer. Food Manuf.* **24**(9): 260–262, 281.

Attaway, J. A. (1994). "Citrus juice flavonoids with anticarcinogenic and antitumor properties." In *Food Phytochemicals for Cancer Prevention I*, ACS Symp. Ser. 546, M. T. Huang, T. Osawa, C. T. Ho, and R. T. Rosen, Eds., American Chemical Society, Washington, D.C., Chap. 19, pp. 240–248.

Beljanski, M. (1992). Pharmaceutical composition and method of use, U.S. Pat. 5,145,839.

Braddock, R. J. and Kesterson, J. W. (1973). Citrus seed oils, FL Agr. Exp. Sta. Tech. Bull. No. 756, University of Florida, Gainesville, FL.

Cho, S. H., Seo, I. W., Choi, J. D., and Joo, I. S. (1990). Antimicrobial and antioxidant activity of grapefruit seed extract on fishery products, *Bull. Korean Fish. Soc.* **23**(4): 289–296.

Cho, S. H., Seo, I. W., and Lee, K. H. (1993). Prevention of microbial post-harvest injury of fruits and vegetables using grapefruit seed extract, a natural antimicrobial agent, *J. Korean Agr. Chem. Soc.* **36**(4):277–283.

Federal Register (1995). Chemie Research & Manufacturing Co., Inc.; filing of food additive petition, *Fed. Reg.* **60**(51):Mar. 16, 14286.

Foo, C. L., Goh, H. S., Ho, Y., and Seow-Choen, F. (1995). Prospective randomized controlled trial of a micronized flavonidic fraction to reduce bleeding after haemorrhoidectomy, *Br. J. Surg.* **82**(8):1034–1035.

Food Chemical News (1995). False data in food additive petition alleged, *Food Chem. News*, August 14, 15–16.

Frazier, S. F. (1980). Antimicrobial compositions of matter from naturally occurring flavonoid glycosides, U.S. Pat. 4,238,483.

Harich, J. (1997). Antimicrobial grapefruit extract, U.S. Pat. 5,631,001.

Hasegawa, S., Fong, C. H., Miyake, M., and Keithly, J. H. (1996). Limonoid glucosides in orange molasses, *J. Food Sci.* **61**(3):560–561.

Hijiya, H. and Miyake, T. (1997). α-Glycosl hesperidin and its uses, U.S. Pat. 5,627,157.

Horowitz, R. M. (1981). "Flavonoids, mutagens, and citrus." In *Quality of Selected Fruits and Vegetables of North America*, ACS Symp. Ser. 170, R. Teranishi and H. Barrera-Benitez, Eds., American Chemical Society, Washington, D.C., Ch. 5, pp. 43–59.

Ko, G. H., Lee, K. H., and Cho, S. H. (1995). A safety test on grapefruit seed extract, *J. Korean Soc. Food Nutr.* **24**(5):690–694.

Lockett, C. G. (1997). Treatment of sickle cell disease, U.S. Pat. 5,626,884.

Moldowan, M. J. and Moldowan, C. (1985). Composition and method for reducing hangover, U.S. Pat. 4,496,548.

Morgan, D. R., Andersen, D. L., and Hankinson, C. L. (1971). Process of inhibiting staling of milk prior to sterilization, U.S. Pat. 3,615,717.

Nishina, A., Kihara, H., Uchibori, T., and Oi, T. (1991). Antimicrobial substances in DF-100, extract of grapefruit seeds, *J. Antibact. & Antifungal Agents, Japan* **19**:401–404.

Paul, S. M. (1997). Ascorbic acid compositions providing enhanced human immune system activity, U.S. Pat. 5,626,883.

Raaf, H. (1985). Tooth and mouth care agent, U.S. Pat. 4,559,224.

Roe, D. A. (1990). "Drug-folate interrelationships: historical aspects and current concerns." In *Folic Acid Metabolism in Health and Disease*, M. F. Picciano, E. L. R. Stockstad, and J. F. Gregory, Eds., Wiley-Liss, New York, pp. 277–287.

Sachs, A. (1997). *The Authoritative Guide to Grapefruit Seed Extract*, LifeRhythm, Mendocino, CA.

Sharamon, S. and Baginski, B. J. (1996). *The Healing Power of Grapefruit Seed*, Lotus Light Publications, Twin Lakes, WI.

Schirra, M. and Mulas, M. (1995). Improving storability of "Tarocco" oranges by postharvest hot-dip fungicide treatments, *Postharvest Biol. Technol.* **6**:129–138.

Sundheim, G. and Langsrud, S. (1995). "Natural and acquired resistance of bacteria associated with food processing environments to disinfectant containing an extract from grapefruit seeds." In *International Biodeterioration & Biodegradation*, Elsevier, Great Britain, pp. 441–448.

Timmons, D. E. (1950). Citrus canning in Florida: Early history and current statistics. AE Series 50-4, January, Fla. Agr. Exp. Sta., University of Florida, Gainesville, FL.

USDA. (1998). USDA Food Composition Data. http://www.nal.usda.gov/fnic/foodcomp.

Will, R. T. (1916). Some phases of the citrus by-product industry in California, *J. Ind. Eng. Chem.* **8**(1):78–86.

Yee, G. C., Stanley, D. L., Pessa, L. J., Dalla Costa, T., Beltz, S. E., Ruiz, J., and Lewenthal, D. T. (1995). Effect of grapefruit juice on blood cyclosporin concentration, *Lancet* **345**(8955):955–956.

INDEX